Commemoration
Colorado School of Mines History Project

The Colorado School of Mines (CSM) Alumni Association has sponsored the Colorado School of Mines History Project and the resulting book. Without that support, the project could not have been completed.

The current CSM History Committee, which has succeeded in producing this book, was convened in August 2001. Committee membership has changed through time, but all persons who have served deserve sincere gratitude and appreciation.

Dr. John U. Trefny, president of CSM, and the CSM Board of Trustees provided strong support and encouragement.

The CSM History Committee thanks the author, Wilton E. Eckley, for his early commitment and excellent and tireless work in writing this book.

The extraordinary support contributed to the publication of a history of Colorado School of Mines is a source of great pride to the school, the CSM Alumni Association, and the CSM History Committee.

We thank all of those whose generosity helped make this history project possible, in particular the following individuals whose gifts and grant/loans were critical to the history's publication. These levels of generosity recognize individuals whose support has been as follows:

PLATINUM LEVEL

Wallace McGregor
Lawrence A. McPeek
John W. Robinson
Robert J. Weimer

GOLD LEVEL

Harry Campbell
David F. Coolbaugh
John D. Haun
Norman R. Zehr

SILVER LEVEL

Hugh W. Evans
Weldon G. Frost
Cooper B. Land

COPPER LEVEL

Charles A. Champion
David R. Cole
Douglas R. Cook
Jack J. Grynberg
Ian H. Mackay
Michael B. Watson

GIFT LEVELS:

Platinum: $5,000 and above
Gold: $2,000 to $4,999
Silver: $1,000 to $1,999
Copper: $100 to $999

CSM Student Life is also thanked for its financial support.

Any net proceeds resulting from the sale of the books will benefit the work of the CSM Alumni Association, as determined by the Board of Directors of the Alumni Association.

*Planning and Editorial Committee:
Seated, left to right—Carol Chapman; Dr. David Coolbaugh '43, '47, '61; Kathleen Feighny. Standing, left to right—
Dr. Robert Weimer; David Powers; Kathleen Breit; Dr. Frank Hadsell '61; Norman Zehr '52, '56.
Not Pictured: Dr. Barbara Olds and Michael Watson.*

THE
DONNING COMPANY
PUBLISHERS

Rocky Mountains to the World

A HISTORY OF THE COLORADO SCHOOL OF MINES

Wilton Eckley

Copyright © 2004 by Colorado School of Mines Alumni Association

All rights reserved, including the right to reproduce this work in any form whatsoever without permission in writing from the publisher, except for brief passages in connection with a review. For information, please write:

> THE DONNING COMPANY PUBLISHERS
> 184 Business Park Drive, Suite 206
> Virginia Beach, VA 23462
>
> Steve Mull, *General Manager*
> Barbara B. Buchanan, *Office Manager*
> Pamela Koch, *Editor*
> Andrea L. W. Eisenberger, *Graphic Designer*
> Stephanie Danko, *Imaging Artist*
> Mary Ellen Wheeler, *Proofreader*
> Anne Cordray, *Project Research Coordinator*
> Scott Rule, *Director of Marketing*
> Travis Gallup, *Marketing Coordinator*
>
> **Barbara Bolton**, *Project Director*

Library of Congress Cataloging-in-Publication Data

Eckley, Wilton.

    Rocky Mountains to the world : history of the Colorado School of Mines / Wilton Eckley.

        p. cm.

    Includes bibliographical references and index.

    ISBN 1-57864-265-5 (hardcover : alk. paper)

1. Colorado School of Mines—History. I. Colorado School of Mines. II. Title.
TN210.C585 2004
622'.071'178884—dc22

                              2004006275

Printed in the United States of America at Walsworth Publishing Company

TABLE OF Contents

Author's Acknowledgments 6

Prologue 7

Chronology 9

CHAPTER 1 From Vision to Reality 12

CHAPTER 2 Beyond Silver and Gold 30

CHAPTER 3 Progress Amidst Adversity 48

CHAPTER 4 Steadying the Course 70

CHAPTER 5 Reaching Out 94

CHAPTER 6 A Decade of Change 116

CHAPTER 7 Still Achieving, Still Pursuing 138

CHAPTER 8 Mines at "100" 160

CHAPTER 9 Prospects 184

Epilogue 206

Bibliographical Sources 218

Index 220

About the Author 224

AUTHOR'S
Acknowledgments

Writing can be a lonely experience, but in this instance I have not found it so, thanks to the Mines Alumni Association, who sponsored this work, and the Planning and Editorial Committee, who took me under their collective wing, providing not only material support but also timely and valuable editorial advice. Members of that committee include Dr. David Coolbaugh '43, Chair; Kathy Breit, Interim Director of the Alumni Association; Carol Chapman, Special Assistant to the President; Kathleen Feighny, Administrator, Economics; Dr. Frank Hadsell, Professor Emeritus (Geophysics); Dr. Barbara Olds, Associate Vice President of Academic Affairs; David Powers, former Vice President of Institutional Advancement; Michael Watson, former Director of the Mines Alumni Association; Dr. Robert Weimer, Professor Emeritus (Geology); and Norman Zehr '52, former Director of the Mines Alumni Association.

I also owe much to the following: Robert Sorgenfrei, Librarian and Archivist at the Arthur Lakes Library, who with much patience guided me through the plethora of material in the library in general and in the Russell L. and Lyn Wood Mining History Archive; Dr. Anton Pegis, Professor Emeritus of English, and Dr. John Haun, Professor Emeritus of Geology, for their valuable editorial advice; the student EPICS team of Kira Kent, Earl Benson, Rachel Biller, Tom McMulkin, and David Wood for their social/pictorial history of Mines; Professor Emeritus Donald Dickinson for photographic aid; Douglas Baldwin '03, graduate student in mathematics, for his artistic skills with the camera and for providing access to his immense collection of photographs; Maureen Keller, Editor of *Mines Magazine*, for her helpful advice in a number of matters; Genevieve Goulding, a work-study student, whose scanning and photographic skills were most helpful as the project neared its conclusion; my son Tim for proofreading; Erika Wilkins for the index; and the many administrators, staff, faculty members, alumni, and friends who shared with me historical material and many interesting stories and insightful information.

Lastly, I owe special thanks to Kaitlin Schmidt '06, the youngest contributor to this effort, who, as a work-study student for more than three semesters and a summer, performed innumerable detailed research tasks in both textual and photographic areas—as well as teaching me some of the finer points of scanning. Her organizational skills, along with her calming sense of humor, helped immensely in moving the project along.

Photographs are courtesy of Douglas Baldwin; David L. Cornwell; Genevieve Goulding; John McMillin (dust jacket); the Pioneer Museum (Golden, Colorado); the Alumni Association (Colorado School of Mines); *Mines Magazine* (Colorado School of Mines alumni publication); the *Oredigger* (Colorado School of Mines student newspaper); the *Prospector* (Colorado School of Mines yearbook); the Russell and Lyn Wood Mining Archive of the Arthur Lakes Library (Colorado School of Mines); and Time-Life, Incorporated.

Prologue

*If the touchstone of a work of art is its precision,
then the touchstone of a university is its history.*

While the present age may be largely marked by a sense of immediacy and self-interest, it is also an age fascinated with history. The human mind has always desired some explanation of the past and how it helps to explain the present and predict the future, if for no other reason than to be able to visualize some order in the course of events—to be able to say, "Yes, that is the way it must have been," even if in reality it may not have been exactly so. History, however, is not a constant but a continuum of change. Each age, therefore, must indeed write its own books and formulate its own interpretations of the past. The Irish novelist James Joyce tells us in his enigmatic work *Finnegans Wake* that "every telling has a taling." Interestingly enough, eight years before the appearance of Joyce's work, J. Harlan Johnson, then an associate professor of geology at Mines, presented a paper on history at the Monday Evening Club in Golden, Colorado (see *Mines Magazine* March 1931). In that paper, he made the following insightful observation: "Most history has been written in narrative form—a form of literature. Therein, perhaps, lies much of the forceful appeal of history; for so long as men endure, the tale of what men have done and how they have striven will never lose its interest and attraction." That, in an important sense, is what history is all about—the telling of stories. And, certainly, the Colorado School of Mines has a story to be told.

Institutions of higher education surely do not have to justify their role in society. Educating youth to participate in the advancement of civilization speaks for itself as a noble endeavor. But the founding and nourishing of a college or university is, to say the least, a daunting task, one marked by challenges to meet, opportunities to grasp, and pitfalls to avoid. In a very real sense, as harsh as it might seem, survival of the fittest is really the hard and fast "bottom line" in higher education. It has been said that the most difficult thing in the world to kill is a college or university. Perhaps so. But even if simple survival is not the question, the fact remains that, to borrow and slightly change a phrase, it is the fittest that "get there fustest and do the mostest." In that respect, Mines has more than earned its international reputation as a school whose graduates indeed rank among the fittest, as well as among those who do the most.

Telling the Mines story, however, is more than simply singing the praises of the school. It is rather presenting as accurate and objective a picture as possible of the institution from its humble beginnings in the foothills of the Rocky Mountains to the place of respect that it holds in the world today. In the game of higher education, every institution, as it should, attempts to present itself in the best light possible. But no single school has the prettiest campus, the best academic program, the brightest students, the strongest faculty, the best athletic program, or the "biggest bang for the buck." Each has had, and will continue to have, its own problems and its own achievements—in short, its own story.

Clearly, this book cannot present a complete history of Mines. What it can do, hopefully, is present a narrative of events with enough significant details to make the story worthwhile to the reader. While the primary focus is on the academic side of the school, attention is also given to cocurricular activities and traditions, alumni activities, and community relationships. After all, Mines may truly be more than a sum of its parts, but those parts, taken individually, have their own importance and their own interest to students, alumni, and friends alike as they contribute to that intangible yet real *Spirit of Mines*.

Editorial Notes

While the school has had a number of name variants—School of Mines, Territorial School of Mines, State School of Mines, and, finally, the Colorado School of Mines—I have chosen, with a few exceptions, to use simply Mines. The alumni magazine has also had variants in its name, with the *Mines Magazine* the most recognized and used. I have, therefore, used the abbreviation MM when quoting or referring to information from that magazine. The present *Golden Transcript* was originally called the *Colorado Transcript*, and I have chosen to simplify both titles to the *Transcript*. The term Archive refers to the Russell L. and Lyn Wood Mining History Archive located in the Arthur Lakes Library on the Mines campus.

Chronology

1859 Gold discovered in Colorado.

1866 Bishop George Randall arrives in Denver as Episcopal missionary bishop of Colorado Territory.

1869 Randall begins construction of Jarvis Hall, a preparatory school, and Matthews Hall, a divinity school.

1872 Jarvis Hall and Matthews Hall open for classes. Mining school under construction.

1873 Mining school opens for classes.

1874 Mining school becomes Territorial School of Mines. E. J. Mallett appointed professor-in-charge.

1878 Jarvis Hall and Matthews Hall destroyed by fire. Mines moved to downtown Golden. Milton Moss appointed professor-in-charge.

1880 First classes held in new building on the present campus. Albert C. Hale appointed professor-in-charge and then president.

1882 First addition to original building.

1883 First commencement. Degrees granted to Walter Wiley and William Middleton. Regis Chauvenet appointed president.

1888 Football comes to Mines. Mines wins state championship from 1889 to 1903.

1890 Second addition to original building (including library, auditorium, and gymnasium).

1892 Elimination of courses for special students.

1894 Hall of Engineering completed.

1895 Alumni Association founded.

1898 Florence Caldwell is the first woman to receive Mines degree.

1900 Winfield Stratton gives Mines $25,000. Assay Laboratory completed.

1902 Charles S. Palmer appointed president. Student unrest. Victor C. Alderson appointed president.

1904 Stratton Hall completed. Mines begins four-year run of Rocky Mountain Conference football championships.

1906 Guggenheim Hall, built with money from Simon Guggenheim, completed.

1908 "M" completed.

1910 Experimental Plan completed.

Rocky Mountains to the World: A History of the Colorado School of Mines

1911 Faculty unrest.

1912 Engineers Company of the Colorado National Guard offers training to interested Mines students.

1913 William G. Haldane appointed acting president.

1915 William Battle Phillips appointed president.

1917 Student unrest. Howard C. Parmalee appointed president. Victor C. Alderson appointed president.

1918 Mines students drafted into Student Army Training Corps. First Battalion, Colorado Engineers expanded into 115th Regiment, U.S. Engineers.

1919 Mines granted permission to organize Senior Engineer Officers Training Corps.

1921 Break between Mines and Alumni Association. Mines investigated by General Assembly. Mines offers first doctoral degrees. Edgar Mine acquired.

1924 First Homecoming celebration.

1925 Melville F. Coolbaugh appointed president.

1927 First Engineering Day.

1928 Mines and Alumni Association found Colorado School of Mines Foundation.

1931 "M" electrified.

1934 First silver diplomas granted.

1937 Field House completed (named for Frederick Steinhauer in 1949). Berthoud Hall completed.

1939 Mines football team undefeated.

1942 Student unrest.

1943 Mines begins quartering and training a basic engineering detachment of the U.S. Army.

1946 Ben H. Parker appointed president—first Mines alumnus to be appointed such.

1948 Colorado School of Mines Research Foundation founded.

1949 Geophysics degree offered. Petroleum divided into two departments—production and refining.

1950 John Vanderwilt appointed president.

1953 Alderson Hall (Petroleum) completed. Dean of Graduate School position created.

1954 Bradford Hall (dormitory) completed.

1955 Arthur Lakes Library completed.

1960 Mines re-accredited by North Central Association.

1962 Group of prominent alumni examines educational status of Mines. Degrees offered increased to nine. End of the Gauntlet as a requirement for freshmen.

1963 Orlo Childs appointed president. Meyer Hall (Physics) completed. Four women enrolled.

1964 Former fraternity house converted to first dormitory for women.

1968 Task force of educational experts evaluates Mines. Paul Newman chooses Prospector Queen—Linda Koerber.

1970 ROTC required for freshmen only. Guy T. McBride appointed president.

Chronology

1973 Mines investigated by American Association of University Professors.

1974 Mines celebrates 100th anniversary. Bachelor of Science in Mineral Economics offered. Russell Volk Gymnasium completed.

1977 Effort made to describe educational goals of Mines.

1980 Mines Plan adopted by legislature, giving trustees more flexibility. George R. Brown Hall completed.

1981 Mines written up in *Newsweek*.

1984 George S. Ansell appointed president.

1986 Stratton Hall renovation completed.

1988 Faculty Senate formed. De jure tenure established with new edition of Faculty Handbook. Berthoud Hall renovation completed.

1989 Confidence vote regarding administration held.

1990 Mines ROTC unit becomes part of the University of Colorado ROTC program. Steinhauer Field House renovation completed.

1991 Mines removed from American Association of University Professors' censure list. Renovation of Engineering Hall completed.

1992 Green Center renovation completed.

1994 Alderson Hall renovation completed.

1995 100th anniversary—Alumni Association buries time capsule to be opened in 2045. Coolbaugh Hall renovation completed.

1997 Study of campus culture by Retention and Management Committee.

1998 Hill Hall renovation completed. Metallurgy Department named George S. Ansell Department of Metallurgical Engineering and Materials Science. Theodore Bickert appointed president. Mines endowment reaches $100 million.

2000 President Theodore Bickert retires.

2001 Dr. John U. Trefny appointed president. Center for Technology and Learning Media completed. Mines recognized by the legislature as an Exemplary Institution.

2002 Opening in Abu Dhabi of the Petroleum Institute, developed by Mines under contract with the Abu Dhabi National Oil Company. Second study of campus culture (Banning Report).

2003 State cuts budgets of institutions of higher education. General Research Laboratory completed. Agreement signed with Alumni Association to establish an Office of Alumni Relations on the campus.

CHAPTER 1

From Vision to Reality

*I swear there is no greatness or power that does not emanate
from those of the earth!
I swear there can be no theory of any account, unless it
corroborate the theory of the earth!*
—Walt Whitman

From its very beginning, American history has reflected both a physical and a metaphysical relationship between man and nature. America's early development, moreover, whether political, cultural, or economic, did not rise solely from the past; on the contrary, in a very real sense it created itself from the reality of the present and the vision of the future and would continue to do so. As it expanded across the continent, America boldly followed a path undergirded and justified by the idea of Manifest Destiny, a concept that, well before it was codified by any specific name, lay at the base of the numerous treks west made by Americans and others of various nationalities. As ranchers, miners, farmers, builders of railroads, teachers, preachers, and founders of towns that eventually grew into cities, these pioneers sought to bring to heel a vast wilderness that was haloed by the promise not only of freedom, but also of wealth and power. From this "westering" and the concept lying beneath it came what has been called the frontier spirit or, sometimes more symbolically, the American Dream. In this broad context of man and nature, of dream and practicality, the story of the Colorado School of Mines begins.

I

Mining of one kind or another has been a part of man's life from prehistoric times to the present, as he has attempted, first with his hands, then with primitive tools, and finally with mechanisms of high technology, to procure from the earth those materials and resources that would provide him not only a sustenance, but also a sense of achievement and gratitude. Through the last half of the nineteenth century and the early part of the twentieth in America, mining was a key activity in the development of the lands west of the Platte River. The discovery of gold in California in 1848 drew thousands of Americans and foreigners alike to try their luck at becoming rich overnight. Indeed, mining became the largest nonagricultural source of jobs in

Opposite page: In search of shining gold.

the West. Ten years later, gold was found in Clear Creek Canyon near what is now Idaho Springs by John Gregory and George Jackson; Golden (then called Golden City) became an access point for the surrounding gold fields. Not exactly a mining town in the sense of a Central City or a Cripple Creek or a Deadwood, Golden was nevertheless a vital link to the gold fields. While Colorado generally, and Golden specifically, did not experience the same degree of frenzy that California did, there certainly was no shortage of hopefuls streaming into the state and through Golden to the nearby mining areas. Colorado was surely enough on the map for mining of gold and silver, as it would be later for mining of coal and uranium and drilling for oil.

Before long, however, outcroppings of metal were becoming scarce, as were the pick-and-pan prospectors. Some of these did on occasion "strike it rich," but for most the dream of wealth ended up a nightmare of backbreaking work under primitive living conditions that yielded little, if any, monetary rewards. Many, having seen enough of the proverbial elephant, left such appropriately named mining localities as Poverty Hill, Poor Man's Creek, Last Chance, or Miller's Defeat and headed home—or wherever. A mining ballad, no doubt sung in numerous mining camps of the West, sums up succinctly in the first stanza more than one sorry prospector's plight:

> It's four long years since I reached this land,
> In search of gold among the rocks and sand;
> And yet I'm poor when the truth is told.
> *I'm a lousy miner,*
> *I'm a lousy miner in search of shining gold.*

The same picture of the hardship of mining in the West, this time in prose, was captured in a letter written by H. A. Johnson from the Nevada City mining camp of Colorado to an Illinois newspaper editor:

> But here, as everywhere else in a mining country, mere surface men and surface work can never succeed, and I am glad that it is a discriminating country. All who come here are not deserving of success. A man must have capital, work hard, and be exceedingly lucky, or come back short of success. I tell you, of all the places I ever saw, this is entirely the worst place for the gentleman of leisure. Hard-fisted, big-hearted, bold, adventurous men may come here, hoping to better their circumstances, while others had better remain where they are, as their board will cost them entirely too much in this country. (MM September 1932)

Mining was caught in the process of change. The days of the romanticized prospector were coming to a close—if they were ever really open. Going underground and demanding more technology and

The Reverend George Randall had a dream of bringing religion and education to the West.

Chapter 1: From Vision to Reality

W. A. H. Loveland was the first president of the original Board of Trustees for the new mining school.

capital, mining would soon be a bonanza of big business and millionaires, who would build opera houses and shoe their horses in silver.

A theme of the popular Western movie genre is that of the heroic teacher or minister, often displaying as teacher a pretty young woman and as minister a former gunfighter trying to redeem his past, coming into a mining town or cattle town to educate the minds and save the souls of the inhabitants. Such, however, was not the case with Golden, whose unlikely white knight, as it were, was the Reverend George Randall, a true Episcopalian missionary bishop from Boston. A graduate of Brown University and the General Theological Seminary of New York, Randall was one of a number of such missionary ministers that various denominations sent to "civilize" the West. Many in the East had ideas not only of "taming" the West, but also of making money from it; and to them, a regulated civil society with all of its amenities was a goal worth supporting, as were men like Randall, who came to plant and nourish both religion and education in a restless, exuberant, and often coarse land.

Arriving in Denver in June 1866 to become one of the few clergy in the Colorado Territory, the bishop found his new surroundings quite different from those he had just left back East. Here was no land of mature metropolises of brick and culture or of large churches and noted universities; on the contrary, here was a land of rawness and excitement, of optimism and risk—where liquor flowed freely and horse thieves were shot. This was the West writ large. Recognizing it as such, Randall was more than ready to put his own idealistic mark on it in terms of religion and education. One of Randall's goals was to build a quality preparatory school to be called St. John's of the Wilderness—a school that would focus not only on the basic arts and sciences, but also on the industries of the territory. While not called St. John's of the Wilderness, the school itself would eventually be built.

First, however, Randall directed his attention to Denver where he founded Wolfe Hall, an Episcopal high school for girls offering a classical curriculum with such courses as languages, literature, music, and painting. Wolfe Hall proved successful enough that Randall could explore possibilities for his original idea of a preparatory school. With a gift of twelve acres of land near Golden from Charles C. Welch, he began this second project, a kind of university that was to have three components: Jarvis Hall, a school to prepare men and boys for college; Matthews Hall, a divinity school; and a school of mines. To some, a school of mines might seem to have little connection with the first two schools. Randall, however, like many others, was aware that if the territory were to have any worthwhile gain from its mining endeavors, it would need a cadre of individuals trained at one level or another in the essentials of successful mining. Den Galbraith in his wide-ranging and informative history "Give 'em Hell, Mines!" phrased the need of the time:

> Finally it dawned on mine owners. Engineers were needed. Bearded, leapfrogging miners had quickly taken up the virgin gravels, then

Edward L. Berthoud

Born in Geneva, Switzerland, in 1830, Edward L. Berthoud came to New York City with his family in 1832. Graduating from Union College in 1846 with a concentration in scientific studies, he gravitated to the field of engineering. After a year working in Panama, he returned to the United States as an engineer, wandering through Kentucky, Ohio, Indiana, Wisconsin, and Iowa. His westward movement brought him to Leavenworth, Kansas, where he spent five years. The lure of gold led him in 1860 to Golden, Colorado.

After early efforts at mining, Berthoud became interested in exploring various areas around the head of Cedar Creek. His discovery of a pass leading into Middle Park led the Central Overland and California Express Company to hire him to survey a route to Provo, Utah. In 1862, he received a commission as first lieutenant in the Twentieth Colorado Volunteer Infantry and ultimately was appointed assistant adjutant general of Missouri counties that bordered Kansas—a hot spot in the times leading up to the Civil War. Promoted to captain in July 1864, he saw action in the protection of Jefferson City, the capital of Missouri. Following various assignments, he was mustered out of service in 1866. Returning to Golden, he turned his energies to engineering and railroad construction, eventually becoming chief engineer and secretary of the Colorado Railroad Company. He participated in building a number of rail lines in Colorado, Idaho, Montana, Utah, and Wyoming.

Along with the above activities, Berthoud worked for the improvement of education, especially with regard to the Colorado School of Mines. His efforts on the school's behalf were instrumental in gaining territorial and state support and, more importantly, for keeping the school in Golden. He served on the Mines Board of Trustees during the crucial early years. Berthoud Hall on the Mines campus is named in his honor.

Pictured above: Edward L. Berthoud, an explorer and railroad builder in the West, was one of the original supporters of the new mining school.

miners began to comb the hillsides. Free gold in the outcrops made a few quick fortunes, but complex sulphide ores could not be treated successfully. Cousin Jack miners [Cornish coal miners] showed them how to mine, but about three-fourths of the metal value ended up in the tailings.

It would be a prim professor from Brown University, Nathaniel P. Hill, who didn't drink, smoke, or swear, who created a breakthrough when he fired up his Boston and Colorado Smelter at Blackhawk in 1868. Hill proved a point—trained engineers could help save Colorado's mining industry. (MM February 1974)

In his comments at the cornerstone laying for Jarvis Hall, Randall noted the need for both religion and learning in the Colorado Territory as it moved toward statehood and praised Golden as a convenient, healthy, and beautiful location for the new schools. However, as anyone who has spent much time on the front range of the Rockies knows, the winds that often roar down from the high country can at times be unkindly, a fact that Randall learned firsthand on 17 November 1869 when one of the strongest windstorms the area had seen in years blew off the roof of the newly constructed Jarvis Hall and collapsed its walls. Apparently, the contractor, or his workmen, did not think it necessary to board up the openings where windows and doors were to be placed, with the result that the wind had free rein throughout the building. This event disappointed not only Randall, but also the general population of Golden. Feeling, as he put it, "a slight touch of Job's emotions," Randall was not to be denied his goal and soon gathered support for rebuilding Jarvis Hall. With a gift of additional funds from the original benefactor, George Jarvis, along with a bell from Mrs. Jarvis and a library of 1,500 volumes from the Reverend Ethan Allen of Baltimore, he moved quickly to reconstruct the destroyed building.

The cornerstone laying for Randall's new building on 8 August 1870 attracted an even larger crowd than did the first, a crowd of opposites—educators and miners, churchgoers and gamblers, housewives and dance hall girls—surely a true crowd of the West. Speaking to one and all, Randall, at the beginning of his address to the huge gathering, referred to the disaster that struck the first structure with a sense of irony that surely no one thought that, before the first edifice should be occupied, the cornerstone of a second would be laid. Giving thanks to both the deity and the territorial legislature, Randall stressed the importance of education in general to the nation and in particular to the West. He also noted the importance of a mining school, one within easy reach of some of the largest mines yet developed and those for ages to come. He closed his address with both hope and confidence that the people of the territory would show their appreciation with a liberal patronage that, together with God's blessing, would enable Jarvis Hall to grow and thrive as a university. Both the hope and confidence that marked his concluding words were indeed to be fulfilled, though perhaps not in the way he would have envisioned. Jarvis Hall and Matthews Hall would pass from the scene, and it would be the mining school of the

This view of the original campus shows Jarvis Hall on the left and the mining school on the right. Matthews Hall, not shown, was to the left of Jarvis Hall.

When the mining school moved to Golden, this building was the first to be built on the new campus, in 1879. (PIONEER MUSEUM)

"university" that would grow to compete with some of the best engineering institutions of higher learning in the world.

Recognizing the efforts of Randall to provide Golden and the surrounding area with educational opportunities and also recognizing the need for a school of mines, the Territorial Assembly, in early 1870, appropriated $3,872.45 for a building to house the proposed mining school. Curiously, some four months prior to the second laying of a cornerstone for Jarvis Hall, the *Central City Register*, evidently feeling that any mining school should be in Central City where there were a number of mines, registered a complaint regarding the $3,872.45 appropriation that the Assembly passed for a mining school as part of Randall's educational enterprise. *The Rocky Mountain News* responded with the following:

> The *Central City Register* does not like the law appropriating $3,872 for the establishment of a school of mines in connection with Jarvis Hall at Golden City. It even goes so far as to declare it a "deliberate swindle." And its only objection so far as we can make out is that the school is not located in a mining town. This does not seem to us a serious objection.
>
> Students do not need to go down a shaft to learn all that is necessary about the sciences of mining and metallurgy, and if an inspection of actual work in the mines would be useful, it will be but a pleasant trip to Central, and a couple of days there will teach them all there is peculiar and important about it. (21 March 1870)

Although the mining building was still under construction when Jarvis Hall and Matthews Hall opened in the fall of 1872, the future school of mines did offer courses under the auspices of Jarvis Hall in mathematics, geology, civil engineering, military science, and Spanish. One year later, with its building still not fully completed but with an increased number of courses and a larger faculty, the School of Mines, though listed under the Department of Physics, opened officially and began its own academic journey. In an unusual trinity of diversity, a divinity school, a preparatory school, and a school of mines carried on their specific educational tasks essentially under one administrative roof.

Unfortunately, and ironically enough, as the life of his "university" was beginning, that of Bishop Randall was ending. He died on 28 September 1873. Though his time in Colorado was short, and though he was not to see the fruition of his endeavors, his vision has endured, clear and challenging, as a cornerstone for all that the newly formed School of Mines was to become.

II

Like Bishop Randall and the thousands of others who made their way to settle and build the West, the new mining school, in its own right, was surely a pioneer, with a challenging and rewarding journey lying before it. Although the Territorial Assembly may have accepted the reality that a mining school was sorely needed, it was not yet ready to provide funds for maintaining the newborn one in Golden, leaving this rather daunting task to the Episcopal Church and anyone else who might wish to contribute. Probably realizing that the mining school had the best chance of long-term survival, both the church and the city of Golden feared that it might fall victim to its own financial situation, as well as to the rising political pressure exerted on the Assembly by Denver to have the school relocated in that city. Golden had already lost the honor of being territorial capital to Denver, an honor that it had enjoyed from 1864 to 1867. The citizens of Golden, after all, looked upon the mining school as theirs, and they were not about to let it slip away. So the debate was enjoined, particularly in the newspapers.

Journalistically, George West, founder and editor of the *Colorado Transcript*, carried the flag for keeping the school in Golden. In an editorial, he made a clear statement directed to Levi Harsh, a member of the Territorial Assembly: "We trust that our immediate representative in the Territorial Assembly—Hon. Levi Harsh—will give a portion of his time

Military Drill
From the *Jarvis Hall Record*, January 1871

In a recent article on Jarvis Hall, the *Golden City Transcript* thus speaks of the military feature that has been adopted to interest and benefit students:

"We regard this as one of the most important features of the school, giving the boys a healthful exercise, improving them both in carriage and discipline. As military drill is a prominent feature of the school, we should be glad to see the students dress in uniform. A cheap, serviceable one can be adopted that will cost no more, if as much, as ordinary clothing. For instance, for the summer, a blue jacket, white pants, and blue cap would be pretty and serviceable, and for the winter, the same jacket and coat with light blue or grey pants. We throw out these hints for the consideration of the Bishop, teachers, and parents, and if it is looked upon favorably, we hope to see steps taken to carry it out, and if not deemed feasible, there is no harm done. This noble institution is one that every Coloradoan should be proud of, and especially so the people of Golden City, and do all in their power to make it self-sustaining, by sending their boys to it for their education. There should be at least forty or fifty students in attendance there from this town alone."

Milton Moss was the third professor-in-charge of Mines.

and attention in the interest of the Territorial School of Mines, an institution that seems to labor under the misfortune of being located in Golden—where it ought to be—instead of Denver—where it ought not to be" (14 January 1874). In the days just before the Territorial Assembly was to meet, the journalistic war intensified among the *Rocky Mountain News* and the *World* in Denver and the *Transcript* in Golden. In responding to editorials in those newspapers that argued strongly against any territorial support for the fledgling mining school—at least partly because of the religious tie to Jarvis Hall and Matthews Hall—West spoke out again in the *Transcript* of 21 January 1874, making a cogent case not only for the status quo regarding the location of the school in Golden, but also one for the very existence of mining schools themselves in Colorado:

> The principles upon which mining and metallurgy depend for success are those of the great sciences, and these, when applied in a proper way to technical purposes, can in no case fail to produce good results or to assist in removing the cloud of technical ignorance under which the mining industry of Colorado has too long suffered, producing losses in time, money and confidence, far in excess of the amount of money that a successful school of mines could in any case call for.
>
> If mining schools did not exist, each individual student would have to provide himself—did he wish to study mining or metallurgy—with costly apparatus and appliances, which would place this line of study out of the reach of the great body of students, to say nothing of the inconvenience entailed or the lack of well-qualified teachers—for the latter cannot exist unless there be institutions that call for their services.

To the view expressed by the *World* (undated clipping) that "a wagon road might be worth more to the Territory than a school of mines" and that "schools of mines are undoubtedly good things, but so far wherever they have been tried, they have proved more ornamental than useful," West struck back with a vengeance in the *Transcript*:

> The statement that mining schools "have proved more ornamental than useful" is entirely gratuitous and about as far from the truth as the average editor gets. The mining schools of Sweden, Germany, and England are not failures—neither is Columbia College of New York, the Sheffield Scientific School of New Haven, the school for natural resources in connection with Harvard College, or even the young School of Mines of the State of Missouri. This latter school has the approbation of the Legislature of that State and is highly commended by such an independent and intelligent journal as the *St. Louis Democrat*.

Also coming to the aid of the school and its debated location with their considerable political influence were such men as E. L. Berthoud, Charles Welch (who gave the original tract of land), and W. A. H.

Loveland. As expected, Levi Harsh introduced *An Act to Establish a School of Mines at or near Golden, Jefferson County, Colorado Territory*. This bill also provided that the present school be deeded to the Territory of Colorado for $500. The bill passed; and after a period of some suspense, the School of Mines on 9 February 1874 was given a second and more official birth, this time as a public-supported institution of higher education that before long would have a permanent campus in Golden—just across the street from another budding Colorado institution, the Coors Brewery.

The Territorial Assembly had fairly specific goals in mind for its new school to meet:

- Unite in one collection "type specimens" with analysis attached of all the principal mines (list of 16 minerals).

- Collect and keep a systematic record of weather observation, temperature, snow- and rainfall, and storms.

- Instruct miners and mill men in chemistry, metallurgy, assaying, mining, mining engineering, geology, and mineralogy.

- Aid and assist the farming interest by obtaining data for the systematic calculation and prediction of the water supply of streams.

- Obtain data to measure the effect that the clearing of forests, cultivation of soil, and system of irrigation has on our water supply. By the present wholesale waste, we are threatened in a few years' time with not only the loss of indispensable material for the mines, but also by loss of water supply that cannot fail to seriously cripple agricultural industry.

- Unite under one control and in one building a mass of information that will clearly set forth the mineral wealth of Colorado.

Albert C. Hale, the fourth professor-in-charge, soon became the first president of Mines.

For the time being, things were indeed looking up for all three of the aspiring schools:

> There are several students at Jarvis Hall preparing for admission to the higher Eastern colleges next fall. As a preparatory school, Jarvis Hall ranks second to none, the vicinity of Matthews Hall allowing of a high proportion of teachers as compared with the number of students, thus leaving each teacher free to follow the specialty for which he is best fitted. The Corps of Cadets and the baseball club pertaining to Jarvis Hall are each in a flourishing condition, so that the inducement to healthful exercise is not lacking. The appropriation for the School of Mines having become a matter of fact, we may expect to see the "University Schools of Colorado" attain a high degree of success during the present year. (*Transcript* 19 February 1874)

By now it was quite clear that as mining grew in economic importance in the American West, so too did the need for people trained in the various

Arthur Lakes

In July of 1959, Mines named its new library after Arthur Lakes, one of the more colorful and beloved of Mines professors. Born in England in 1844, Lakes came to Colorado in 1862 to carry out research in field geology. Playing an important role in helping Bishop Randall found Jarvis Hall, Matthews Hall, and the School of Mines, he has sometimes been referred to as "the real founder of Mines." An outstanding field geologist, he became head of the Geology Department at Mines in 1880. A true Renaissance man, he had interests ranging from gathering geological data across Colorado, to drawing and painting, to studying dinosaurs and other prehistoric creatures. He is credited with providing the state with more geological data and surveys of mineral deposits than any other man.

In addition to geology, Lakes taught a freehand drawing course, not only to regular Mines students, but also to many ladies of Golden in special evening classes. In an article for the Mines *Bulletin*, May 1903, he stressed the importance of this skill to the fields of mining and engineering, noting, "how few people can make a fairly accurate freehand sketch in these fields, where from the very nature of the case freehand sketching would appear so essential. Whilst you will not meet one mining engineer in a hundred who is not a fair mechanical draughtsman, you will not meet one in a hundred who can make a fair freehand sketch in the field." While Lakes felt a freehand sketcher, like a poet, must be born, he also believed that "if early education were properly attended in this line as it is in other branches, a large proportion of people who now say they 'cannot draw a line' could make a very fair average sketch and represent their ideas in this quick and forceful manner."

Typically the English gentleman, he had an eccentric turn that added an interesting dimension of absentmindedness to his personality. Folklore has it, for example, that after his wedding to Edith Slater in Denver, he drove back to Golden in his carriage alone. Following his teaching career at Mines, he became a mining engineer, working with the United States Geological Survey and editing several mining publications. His published books include *Geology of the Colorado Coal Fields, Geology of Colorado and Western Ore Deposits, Prospecting for Gold and Silver in North America, Geology of Cripple Creek Ore Deposits*, and *Geology of Western Ore Deposits*. He died at the age of seventy-three years on November 21, 1917, in Nelson, British Columbia.

Pictured above: This sketch shows one of the talents of early faculty member Arthur Lakes.

---

scientific and engineering aspects of that activity—geology, metallurgy, civil engineering, chemistry, and the like. Opportunities for such study in America during this period were scarce at best; and those Americans who did receive training in mining engineering in the 1860s were mostly limited to Columbia School of Mines in New York, established in 1864, or to European schools such as the Bergakademie in Freiberg, Germany. A small number of American universities and individual mining schools, the great majority east of the Missouri River, had programs under development; but in most cases they were questionable at best. Engineering in general still had to fight the battle of respectability as a bona fide profession. The line between engineers and artisans was still blurry and would remain so for some time, as one of America's most noted engineers, Herbert Hoover, clearly recognized some years later in the Preface to Thomas Read's *The Development of Mineral Industry Education in the United States* (1941): "My own impression is that it was not until the American universities embraced engineering as a full part of their work that engineering took on the full dignity of a profession. The major demarcation between an artisan and an engineer is not alone the actual technical knowledge. It is the recognition of the qualifications of broad education and obedience to a code of ethics." Hoover emphasizes the necessity of a broad education for mining and metallurgical engineers: "The production of metals and minerals involves all the physical sciences. Its daily need is economics and sociology. It requires character and breadth in men to which all the humanities must contribute."

At this time, however, engineering had yet to achieve the dignity that Hoover referred to. Indeed, for most of those in the West hoping to find at least their subsistence if not their fortunes in mining, learning on the job was about the only road available to them. Unsurprisingly, there was a fair amount of mistrust on the part of the self-made miners regarding the mining engineers with diplomas—that is to say "book miners." Because mining at any geographical location and at any level of expertise, "book" or otherwise, is a rough and dangerous hands-on occupation and demands those not averse to risking life and limb, miners all over the world have not only a pride in their work, but also a kind of "give 'em hell" attitude that stems from their close physical relationship with the earth. Moreover, their respect for each other is strongly based on job performance, whatever the job may be.

III

In light of the status of mining education at the time, we could well say that the fledgling Territorial School of Mines was born, or reborn, in Henry David Thoreau's proverbial "nick of time." As required by the 1874 act passed by the Territorial Assembly formally establishing the Territorial School of Mines, the group of trustees that the Assembly appointed met on 6 July 1874 to organize themselves into a regular board. W. A. H. Loveland was chosen president and Edward L. Berthoud, secretary and registrar. The remaining members were N. P. Hill of Black Hawk, Alpheus Wright of Boulder, W. W. Ware of Georgetown, C. C. Davis of the United States Mint in Denver, J. Marshall Paul of Fairplay, and William Amsbary of Canon City. The new Board of Trustees began preparations for the opening of the school in September 1874, striving to appoint a faculty, organize space, and obtain necessary equipment and materials. The faculty included Professor-in-Charge E. J. Mallett, Theoretical Metallurgy and Chemistry; E. L. Berthoud, Civil Engineering and Geology; Richard Pearce, Theoretical and Practical Metallurgy; Julius Pohle, Mechanical Preparation of Ores; Francis Schmidt, Mathematics, German, and Spanish; Arthur Lakes, Freehand and Mechanical Drawing; and William West, Blowpipe Analysis.

Perhaps in some respects an unseemly looking group, these young men were among the early Mines students.
(PIONEER MUSEUM)

Captain James T. Smith was one of Mines' most fervent supporters, serving forty-six years on the Board of Trustees.

The curriculum was based on a three-year course culminating in a diploma, but students were allowed to enroll under the category of special student for almost any kind of program they desired. Because most of the enrollment was comprised of special students, the instruction was strongly slanted toward the practical in order to prepare them to apply immediately their knowledge of mining and metallurgy. Special students, then, were more or less transient, with the result that planning curricular offerings by the faculty was truly complicated. This was a situation that the school would have to contend with for some years to come. But at the time, a student was a student, so the more the merrier. While some textbooks were available, much of the information was in lecture form, requiring copious note taking. An interesting addition to the regular course offerings was the opportunity for students and faculty to participate in field trips to view geographical and geological points of interest. Golden citizens often accompanied them on these trips. Tuition was $55 for each of the first two terms and $40 for the third term. While $150 might seem quite cheap in comparison to today's expenses, it was a considerable amount at the time and was borne entirely by the student. Special courses required additional fees as follows: analytical chemistry, $135; assaying and blowpipe analysis, $40; photography, $25; and $10 to $15 per course for special students. Room, board, and laundry were available at Matthews Hall for $8 per week.

Two significant items of good news came at the end of the first year of the school's new status. First was the 26 to 4 baseball victory over Central City. Second, not so dramatic but even more important, was that the school was able to complete its first year in the black. Territorial appropriations for the year were $5,000, with tuition and assay receipts totaling $179.43. Expenditures were $5,100.15, thus leaving a balance in hand of $79.28. Would that any college or university of today could be so exacting in its budgeting.

With the departure of Professor-in-Charge E. J. Mallett, who hoped to set up his own mining school in Denver, Professor Gregory Board, a graduate of the Royal School of Mines in London, England, arrived to fill the position. He was welcomed with a glowing article regarding his education and background in the *Transcript* of 17 March 1875 that closed with an interesting comment: "So far as education is concerned, therefore, he comes well furnished to occupy his responsible position, and from what we have seen of the man, we imagine he will not be more nice than wise but will take hold of the work that comes to hand with a full determination of doing it promptly and well." During Board's two years as professor-in-charge, Mines was still struggling, with enrollment hovering around a dozen, counting both regular and special students, and with funding still a problem. The school, however, was beginning to gain notice and received a medal from the Centennial Exposition in Philadelphia in 1876 for the best geological collection on display.

Enrollment, unfortunately, was still only about a dozen, including both regular and special students, for the 1876–77 academic year and only a few more by 1877–78. The now state of Colorado assumed control of the school from the territory in 1877. On 29 March 1878, Milton Moss succeeded Gregory Board as professor-in-charge when the former took over the Golden Smelting Works. Prior to his appointment, Moss had been studying mining and metallurgical methods as they were being practiced in the West. The *Globe* of 16 February 1878 welcomed Moss as a wise choice for the position of professor-in-charge, one who was totally absorbed in his work. "Yet," the article continued, "with all this it cannot be expected of him to surmount all the immeasurable difficulties which besiege the progress of a young institute." We might wonder what Moss himself thought of this rather deft combination of optimistic praising and pessimistic forecasting. But he did not have much time to contemplate anything.

General George West

Born on November 6, 1826, in Claremont, New Hampshire, George West was a journalist, a pioneer, an entrepreneur, a soldier, one of the founders of Golden, Colorado, and an inveterate champion of the Colorado School of Mines. After a journalistic apprenticeship, West, along with some friends, founded Mechanics Mining and Trading Company in 1859 and headed to the gold fields of Colorado. In the area that was to become Golden, West constructed his Boston Company building and served as the first president of the Golden Town Company. In May of 1862, West was appointed captain of Company H of the Second Regiment of the Colorado Volunteers, a regiment that, as part of the Second Colorado Cavalry, fought against both guerrilla bands and regular Confederate units in Missouri and Arkansas in 1863 and 1864. On a leave of absence in 1863, he married Eliza Boyd, who worked in a military hospital at Sedalia, Missouri.

Following his military experience, West returned to Golden to found the *Colorado Transcript* in November of 1866, a newspaper that remained in the West family until 1960 and is now published as the *Golden Transcript*. A Democrat, though he never ran for office, West often engaged in political debate via the printed page. In an article on West appearing in the *Denver Westerners ROUNDUP* of December 1969, Charles S. Ryland recounts one of West's many political squabbles, this one with Lieutenant Governor William G. Smith, editor of the rival *Golden Globe*, when Smith accused West of charging the cost of a saddle horse he used in the 1889 Ute uprising to the state as a war expense.

The two editors fought for a time with the pen, and, finally, Editor Smith decided he had been called a liar and challenged West to a duel. As the one who was challenged, West had the choice of weapons; as he announced in his paper, his choice was "Thundermugs at thirty yards, roll 'em or throw 'em." West's grandson, Neil Kimball, states that his grandfather said that the affair was a "spirituous" [*sic*] hoax "over a bottle of the best."

Governor Alva Adams appointed West as adjutant general of the Colorado Militia in 1887. West was quite active in this position, not only dealing with some Indian uprisings, but also turning the militia into an efficient military unit. According to Ryland, he obtained new uniforms for the militia and was influential in establishing armories in Denver, Leadville, and Colorado Springs. West died on November 15, 1906. Camp George West was named for him on May 1, 1934.

The Golden smelter was a busy place in the town's early days.
(PIONEER MUSEUM)

When the Golden fire alarm rang out around noon on 4 April 1878, the first alarm in many months, it caused considerable excitement in the town. One of its prize structures, Jarvis Hall, was burning through the roof. Militating against the efforts of two companies of firemen was a lack of water with which to combat the fire. Indeed, the scarcity of water had from the time of the founding of the schools caused concern that eventually the three schools might have to be moved to another location. For Jarvis Hall, however, such a question became moot. It was burned to the ground, though Matthews Hall and the mining school, each of which was about fifty feet from Jarvis Hall, were saved. The cause of the fire was attributed to a faulty flue.

Lightning may not strike twice in the same place, but fire surely does. At about midnight of 6 April, just two days after Jarvis Hall was destroyed, the fire alarm sounded once more amid cries that Matthews Hall was ablaze. Starting in the belfry, either from combustion or arson, though there was little evidence of arson, this fire, like the last, made considerable headway before it was discovered; and Matthews Hall too was a total loss, leaving Mines standing alone beside two huge piles of still smoldering rubble.

Not long after these fires, Moss was faced with another and perhaps even greater challenge, that of merging with the State University in Boulder. The basic argument here was whether the educational and economic advantages lay with consolidation or separation, a common argument in American educational history. James T. Smith, in his "Early History of the Colorado School of Mines" (reprinted in *Colorado School of Mines Quarterly*, No. 1, 2000), noted that the "location of the school at Golden, by the state constitution, served to stay the agitation in the General Assembly; and, at the urgent request of its local friends, the school was given a further chance to make good." Smith further commented on what he called the "intimate" feeling between Boulder and Golden: "It is, however, wearing away, the separate location theory, in which the builders built wiser than they knew, having been long since justified by the noted success of the greatest school of mines in the world." Smith's extravagant prose no doubt stemmed from his longtime membership (forty-six consecutive years) on the Board of Trustees of Mines and from his service to the school in numerous ways over the years.

With the loss of the support system provided by both Jarvis Hall and Matthews Hall, which, given the water problem, was no doubt a blessing in disguise, Mines vacated its building and moved its classes to the former Territorial Capitol building in downtown Golden. It was then that the citizens of Golden, recognizing the plight of the school, purchased and donated to the Board of Trustees a very desirable plot of ground 150 by 150 feet. The Colorado General Assembly also came to the aid of the now homeless school with a tax levy for its support. The resulting funds, along with the gift of land from Golden, enabled the Board of Trustees to appropriate $7,500 for a new building, a structure that was the first building on the present campus and turned out to be the south third of what eventually would become the Hall of Chemistry. The new building housed its first classes in October 1880, one month after the appointment of Albert C. Hale as professor-in-charge, a title changed almost immediately to president, with a salary of $2,500. The new structure was described by the *Denver Republican*:

> The building is brick, manufactured from the excellent clay of the neighborhood. Its sightly location, fine architectural style and proportion give it a very pleasing effect while its windows command views in every direction of surpassing beauty and grandeur. The plan of its interior was the result of careful study on the part of the Board of Trustees, under advice of the Faculty, and, in points of convenience and adaptation to the purpose in view, is well nigh perfect. Two stories in height, the front on each floor is devoted to halls, offices, and recitation rooms. The main part below affords a spacious and most convenient chemical laboratory, besides three rooms in the rear occupied by the Assay Department. The corresponding space upon the upper floor constitutes a large and beautiful lecture room, the center being seated, sides and one end occupied by handsome cases containing the various cabinets, while the remaining end furnishes room for the lecture stand, fitted with "every modern convenience." (undated clipping in Archive)

The article praised the faculty and the curriculum and stressed the importance of the mission of the school to Colorado, pointing out that if "knowledge is not power itself, it is the great economizer of power in all industrial pursuits."

Like Mines, the Coors Brewery has played a leading role in the history of Golden. (PIONEER MUSEUM)

*My sweetheart's the mule in the mines.
I drive her without any lines.
On the bumper I sit, and tobacco I spit
All over my sweetheart's behind.*

IV

Mines, though with some distance still to go, was moving toward the status of a full-fledged institution of higher education, soon to grant its first degrees. The *Daily Democrat* reported on 11 May 1881 that the school would prove successful beyond expectations and that there would no longer be a need "to look beyond the borders of our own state for those advantages of education in all departments connected with our chief industry, which have long been the element most lacking and necessary to the fullest and most successful development of our unparalleled resources." What Bishop Randall envisioned and worked for was indeed coming to pass.

According to the catalog of 1880–81, the school offered three three-year courses leading to the degrees of Engineer of Mines, Bachelor of Philosophy in Metallurgy, and Civil Engineering. Hale's report of 1880–1882 detailed the constituency of the student body. A total of 69 students attended the school, including 18 regular students and 51 of the ubiquitous special students. Of the latter, 13 were women attending lectures and drawing classes taught by the popular Arthur Lakes. In addition to Colorado, seven other states were represented in the student body. Eleven students were under the age of 20; 22 were between 20 and 30; 15 were between 30 and 40; and 13 were over 40—bringing the average age of the student body to an elderly 31. Eleven students had degrees from other institutions, including two doctors of medicine and two doctors of dental surgery. Moreover, 22 students said they were prospectors, and 3 said they were superintendents of mines. All in all, a fairly well diversified student body. The 1882–83 catalog shows an increase in enrollment with 37 regular students, 67 special students, and 33 women in the drawing class taught by Arthur Lakes, making a total of 137. With the end of the 1882–83 academic year, the school, in an effort to regularize its course offerings, undertook its first real curriculum revision that eliminated a number of the nondegree courses taken by special students. This move, of course, reduced the number of such students, most of whom rarely stayed for a full school year anyway.

The year 1882 saw the granting of the first honorary degree by Mines, that of Engineer of Mines to Milton Moss. Also in 1882, the Board of Trustees, having appropriated $50 for athletic equipment in 1881, added another $50. These may seem paltry sums, but when the total value of the entire school at that time was about $20,000, they were certainly not to be taken lightly. Even more significant than their amount, however, these appropriations indicated that athletics was on the school's agenda. On a rainy 1 June 1883, the first commencement exercises were held amid flowers and bunting in the lecture hall in Mines' new building. The first earned degrees, Engineer of Mines, were granted to Walter Wiley and William

Chapter 1: From Vision to Reality

Middleton. Hailing the exercises with the headline "The Freiburg of the Rockies Graduates Its First Class," the *Daily News* of 2 June 1883 described a ceremony that was much more intimate than those commencements of today in which hundreds of graduates and guests gather. Following a musical number by a male quartet, both graduates read their essays. Wiley's was titled "The Mining Treatment of Low Grade Ores" and Middleton's, "Legitimate Mining." In his commencement address, President Hale summed up the status of the school:

> Thus far the work done for this institution has nearly all been that of laying a broad and solid foundation. There may be those who have been impatient that the costly and gorgeous superstructure has not risen magic-like from the ground, but in that event, like Jonah's gourd, it would have been destroyed as suddenly as it rose. All permanent growths are slow, and no true friend of the institution will be disappointed because the superstructure has not been erected before the foundation was completed. As for the foundation itself, let judgment be passed upon that only by those who are thoroughly acquainted with what it really is and are capable of saying what it really should be. (*Transcript* 19 February 1874)

For a more descriptive treatment of the commencement, see *Golden: the 19th Century* (1987) by Lorraine Wagenbach and Jo Ann Thistlewood.

As Robert Sorgenfrei noted in his excellent article on the early years of Mines, Hale had the respect of the Board of Trustees, the legislature, and the citizens of Golden, and became increasingly unpopular with the students: "In 1883, the board conducted an investigation after receiving a petition signed by most of the students alleging improper grading practices, extended absences from the school, and unfair treatment. The results of the investigation were kept confidential, but it appears that Hale was cleared of any wrongdoing."

Sorgenfrei added that most Mines students of that time were older than the average college student and had considerable work experience, and as a result "tended to object, sometimes loudly and forcefully, to the kind of paternalistic treatment common on most college campuses of the day" (*Colorado School of Mines Quarterly*, No. 1, 2000). The *Transcript* of 2 May 1883 dismissed the entire incident between Hale and the students as "a veritable tempest in a tea-pot" and concluded, "We have hereto refrained from mentioning the matter at the request of members of the board, and now that the investigation has been had, we find actually nothing to mention." Nevertheless, for whatever reason, Hale resigned his position of president.

With Hale's resignation, the school turned to Regis Chauvenet of St. Louis to become its second president. A graduate of Washington University, Chauvenet received his doctorate from Harvard University. Prior to his coming to Mines, he worked as an analytical chemist in St. Louis.

While the years up to this point may be seen as the formative period of the school, Chauvenet for one did not see them so, stating in his unpublished history of Mines (available in Arthur Lakes Library) that the school "as it opened in the fall of 1883 was certainly in no settled condition, nor was it until fully seven years later." Establishing a definitive demarcation point in the history of any institution is frequently debatable. Although history may be divided into periods for convenience, history is a continuum during which change always leads to more change. Happily, Chauvenet's leadership came at the right time and did have a profound effect in leading the school toward a clearer focus of its mission.

CHAPTER 2

Beyond Silver and Gold

*Now, though the art of husbandry, which I willingly rank
with the art of mining, appears to be divided into many
branches, yet it is not separated into so many parts as this
art of ours, nor can I teach the principles of this
as easily as Columella did of that.*

—Agricola

When Regis Chauvenet assumed the presidency of Mines in 1883, America was anticipating the future, primarily because of what the end of the current century augured for the coming of the next. If the driving force of Manifest Destiny spurred the settlement and development of the American West in the mid-nineteenth century, it would be the forces of industrial and financial enterprise that would mark the later years of that century. Bolstered by social Darwinism (you get what you earn, and you deserve what you get), the entrepreneurial giants driving this new age were perhaps best represented by Andrew Carnegie, Cornelius Vanderbilt, John D. Rockefeller, J. P. Morgan, and Meyer Guggenheim, among others. America, through management and efficiency, supported by the scientific method, was beginning to become a world power in technology, industry, and economics.

American higher education was to play a significant role in the vast and exciting changes that were rapidly overtaking the country. Colleges and universities nationwide were forced to reevaluate and redefine their roles and their strategies in preparing their growing numbers of students to face a world undergoing constant change and to bring a semblance of order to it. The broad spectrum of higher education would be in flux for virtually the whole of the new century in terms of curricula, faculty, equipment, financial support, buildings, and, probably most of all, students, who themselves would face the dilemmas of choosing a school and paying for it and deciding which branch of which profession to follow. Mines, still a relatively small institution with many day-to-day challenges, was methodically preparing its own course for the new century with both hope and wonder.

I

When a faculty member at Mines was interviewing a prospective student in the fall of 1883, the young man said, "I want to quit as soon as I can assay for silver and gold." When the faculty member mentioned some other metals, including zinc, the

Opposite page: Early chemistry laboratory.

Regis Chauvenet guided a fledgling school through its early growing pains.

young man responded in no uncertain terms, "Zinc! No zinc for me. You don't get but three dollars for a zinc assay, and I can do a dozen gold and silvers while one of these laboratory sloopers is doing a single zinc!" Regis Chauvenet recorded this exchange in his history of Mines to illustrate one of the perennial problems the school faced from its very beginning: the quick-in and quick-out mentality reflected in many of its students that made it difficult to establish a viable curriculum, with the result that the school was often referred to during the 1880s as the School of Assaying.

Because of the demand for trained assayers, assaying was one of the most popular course offerings at Mines. Assay offices were, figuratively speaking, on every corner in Denver, and their business was lively. Without assayers, the mining industry would obviously have been in dire straits. Requiring knowledge of chemistry and metallurgy, assaying is a skill, perhaps even an art, that goes back to the ancient Egyptians. Basically, it is the procedure that measures the presence of gold, silver, or some other metal in a sample of ore, ranging from rubbing the ore on a touchstone to heating it to a high degree in a furnace. Georgius Agricola in his *De Re Metallica* (1556) outlined the various techniques of assaying of that time in some detail. A more recent discussion of assaying, however, appears in an article by Roland B. Fischer '42, titled "Ancient Art of Assaying Lives On," which presents a clear and concise history of the procedures of assaying from ancient times to the present, as well as describing the first assaying courses taught at Mines in the 1870s:

> In 1900, an assay building was completed with 16 furnaces. The early assaying courses at Mines were much more than a mental exercise, although learning that part was challenging enough! The coke-fired furnaces were tricky to control properly. Once fired to the right temperature, the furnaces could be run for several days, for instance, during a Christmas or Easter vacation. In the 1940s, gas-fired furnaces were used. Old-timers are quick to point out that this made the course a snap. Not quite so. The newer gas furnaces had a nasty habit of flaming out, and a tossed match for lighting could result in an unnerving explosion in the laboratory. (MM September 1981)

An early focal point on the campus was the president's residence.

Fischer pointed out that a number of early Mines graduates made a career of assaying, including Charles O. Parker '23 of Denver, "who was well-known throughout the West for his accuracy and responsibility." Given the need for assayers and the remuneration that they could command, it is not difficult to understand the quick-in and quick-out strategy of many students, as well as its negative effect on the Mines curriculum. To be sure, Chauvenet understood the problem quickly enough and began to redress it.

Chauvenet soon realized that his greatest challenge was going to be a "struggle between management and environment." The community, he commented in his history, "so far as it troubled itself to have any ideas of the subject, looked upon a technical school as a place to which a student might apply for any line of instruction, irrespective either of the preparation of the applicant or the schedule of the school." His goals in meeting such a challenge were simple and direct: (1) eliminate special courses for special students; (2) expand the curriculum to include basic engineering courses; and (3) strengthen the faculty. These goals may not sound too difficult, but bringing about change in a college or university is somewhat akin to turning the proverbial locomotive around.

The catalogs and previous reports by professors-in-charge and President Hale substantiated Chauvenet's view: "So many were the irregularities, so few the graduates, that true 'course' students were hardly to be found. All that could be done in the very imperfect condition of the courses was to exact good work in the lines made possible by the schedules and to see that the candidate for a degree was of such 'caliber' as to render probable his success in the technical world." This is an important topic in Mines' development over the next decade. It was, as Chauvenet clearly recognized, a pragmatic matter of compromise between the ideal upon which the school was founded and the reality of everyday operation, not an uncommon necessity for collegiate institutions

The Assay Laboratory played a significant role throughout many years of Mines' history. (PIONEER MUSEUM)

or for any other institution. Chauvenet continued, "To have attempted to meet the wishes of every applicant, that is, to be all things to all men, would have resulted in confusion worse confounded from which it would have been futile to essay the organization of a creditable curriculum." Another perennial reality was the lack of funds to permit an immediate all-out attempt at curriculum change or, for that matter, at strengthening the faculty. So, what to do?

First, Chauvenet convinced the faculty (seven members including himself) to introduce a temporary course of study that granted an assay diploma and, at the same time, to put more emphasis on the regular degree course, then four years in length. The positive result of this change was that the regular course became the mode by 1890, making the special student a rare bird indeed. Next, a plan to increase the size of the faculty got a small start with the appointment of Benjamin Sadtler, a Freiburg graduate, to the chair of Metallurgy and Mineralogy in 1887, bringing the total faculty to eight. Not to be overlooked was the arrival with Chauvenet in the fall of 1883 of the inimitable Paul Meyer as professor of Mathematics. A mathematician of considerable international repute, as well as a physician and musician, he was to play one of the more useful and endearing

A Letter Home—
from a Freshman in 1893

The following are excerpts from a letter written by Mines freshman William Mageneau in the fall of 1893 to his mother. The William Mageneau collection of letters is available in the Archive (Lakes Library).

The school has the funniest arrangement I ever saw, and that is what all the new boys say. You come and go as you please, and no one cares or notices. If you miss a lecture, it is your own loss and no one but yourself knows the difference. The professors seem to take it as a matter of course that we knew just how the school was run, and they were too busy to ask them about it. But then, we got there just the same.

They teach almost altogether by lectures though later we will have some recitations. We all have notebooks and record all the important facts in the lecture. This at first seems easy as Professor Chauvenet says the boys think what a snap they will have sitting listening to the teacher recite the lessons for them. But when the exams come, it is an entirely different thing. The exams here last days instead of hours; that is, the examination in a subject, say chemistry, may take two days to complete.

At present I have a great deal of time on my hands, but Mr. Chauvenet told me that he thought I wouldn't have that to complain of later on. I have spent, for school supplies so far, $4.25 for books, that is Geometry, Algebra, and Chemistry, and a trifle for notebooks, etc. For a deposit of $1.00 at school we get a key to a locker in the basement where we can safely leave books, etc. I also paid $1.00 for Gymnasium fee for this term. Other deposits are $1.00 for drawing instruments.

Tell Papa that Mrs. Johnson [boardinghouse owner] gave me $6, which he had paid her. She says that when he asked what his bill was she made a mistake and told him $10 instead of $4. I will keep the $6.

Your affectionate son,
Willie

---

roles as a Mines faculty member and as a medical practitioner in Golden.

Because the Board of Trustees was more interested in erecting a new building on the campus than in increasing the size of the faculty—not an unusual setting of priorities for boards of colleges and universities—Chauvenet's goal of increasing the faculty was put aside, though not forgotten. The building that the trustees were interested in was a metallurgical plant, a facility that might well have been of more value to the mining industry itself than to the education of future mining engineers. Chauvenet, at any rate, objected to such a structure in a brief and precise letter to the board emphasizing

The interior of the Assay Laboratory had some resemblance to Dante's Divine Comedy.

that its contributions would not be commensurate with its construction and operation costs.

Chauvenet had his own plan that he hoped the trustees might be amenable to, one specific of which was an addition to the original academic building that would, as he put it, "confer some dignity upon the School, by including a library room, administration department, auditorium, and other deficiencies in the existing buildings." The school would eventually get an experimental plant, but Chauvenet was correct in his view that Mines sorely needed basic improvements to its campus facilities. The addition was completed in 1890, one year after the completion of a new presidential residence to house Chauvenet and his new wife. In funding the construction of the new academic building, the trustees took what, for a relatively new institution, was a daring step, that of using borrowed money. As Chauvenet quickly responded, the school was never without a financial backer of some kind and always managed to pay its bills on time, though on some occasions borrowed funds may have been used.

The gymnasium, located in the basement of the addition, measured sixty-five feet long by forty feet wide by twenty-three feet high, complete with hot and cold showers. Described as the best equipped of any college or school gymnasium in the state, this facility, according to the Mines Catalog of 1890, included "pulling weights of every description, while among the other instruments are swinging rings, parallel bars, horse buck, quarter circle, 'cage' with apparatus for development of every set of muscles, ladders, springboard, complete set of clubs and dumbbells and many others." Evidently, there were no required physical education courses as yet, but the gymnasium was open every afternoon and three evenings a week for both systematic and free exercising. Students paid one dollar a term for use of the facilities. While some of the instruments of exercise mentioned above might bewilder the visitor of modern fitness centers, it seems that Mines students of the time had the opportunity to be in prime physical condition.

In 1890, Louis C. Hill was appointed to the chair of Physics and Electrical Engineering. The faculty, however, factoring in both the retirements and the additions that had occurred, numbered eight, an increase of only one during Chauvenet's first decade at Mines, hardly what he had hoped for. Although student numbers were increasing steadily, the faculty with its slow growth would take five more years to reach ten. Growth in buildings, however, was not slow. In 1893, the legislature appropriated $20,000 for a physics building, actually called the Hall of Engineering. Completed in 1894, it was, counting the basement, a three-story brick-and-stone structure seventy-five by fifty feet. Featuring

The Hall of Engineering is now the oldest building on the Mines campus.

Though few in number, Mines students were active in many nonacademic ways. (Pioneer Museum)

large, well-lighted rooms for lectures and laboratory work, as well as smaller rooms for special purposes, it housed the physics and electrical engineering departments more than adequately. It is presently the oldest building on the campus.

It was this year that Chauvenet saw as the end of the formative period of Mines, primarily because of the elimination in 1892 of special courses for special students. There was, however, still a problem regarding class enrollment, as he noted in his history:

> It need not surprise us to find in these years (1892 to 1901) great disparity in the numbers of the classes. The preparation of the average freshman was not sufficient. The consequence was a disproportionate freshman class. Not once, from 1892 to 1901, did one half of the freshmen succeed in "making" the sophomore class. The latter class, however, was not so small as might be inferred from this fact, as most of the "advanced" admissions, chiefly from other technical schools, were to sophomore standing.

This lack of balance between the freshman and sophomore classes, along with that of students admitted and those who eventually graduated, would soon present the school with an unexpected challenge.

A signal event occurred in the spring of 1895. Robert Stockton of the graduating class of that year urged his classmates to form an alumni association with membership available to all graduates of Mines. And so, immediately following the commencement ceremonies on 7 June 1895, the Colorado School of Mines Alumni Association (CSMAA) was founded, having elected William B. Middleton '83 president. Mines graduates at the time numbered a mere sixty, of whom twenty-three graduated in 1895. While the association grew in numbers, it grew also in its crucial function through the years as Mines itself expanded and flourished. The relationship between the school and the alumni association, however, was not without a pitfall or two. If ever there were two organizations dependent upon each other and that worked to sustain each other, it was Mines and its alumni association.

II

Significantly, from its inception, Mines was open to all men and women qualified for entrance. With the exception of the numerous women who took the freehand drawing course offered by Arthur Lakes in the early days of the school, Mines, given its purpose, was in its early years essentially a male institution, with all of the characteristics that such connotes. The year 1898, however, with little fanfare, saw the graduation of the first woman from Mines—Florence Caldwell. A native of the small river town of Gallipolis, Ohio, she attended Adelbert College (now part of Case Western Reserve University) in Cleveland and Ohio Wesleyan University, from which she earned a Bachelor of Science degree in 1890. Entering Mines as a sophomore in 1895, she received her degree in Civil Engineering in 1898, becoming one of a very few women of that time to hold such a degree. Although Caldwell chose a career of teaching and homemaking rather than engineering, her graduation was a milestone in Mines history.

Five years passed before the next woman was graduated from Mines. Grace McDermut came from a gold-mining family and attended Mines to prepare for managing the mine, but the family sold the mine before McDermut finished. Following her graduation in 1903, she worked for the National Bureau of Standards. The first woman to be hired by the Bureau in a permanent position and the only woman employee there for five years, she earned a medal for meritorious service in 1949. At her death in 1979, she was the oldest graduate of Mines. Ninetta Davis, entering Mines at the young age of sixteen, was the third woman to earn a degree from Mines, taking her E.M. in 1920. In an interesting career, Davis worked for the Midwest Refining Company in Casper, Wyoming, the Union Oil Company of California in Fort Collins, and the U. S. Geological Survey from 1934 to 1944. After fifteen years with the Shell Oil Company, she started a second career coaching the Dutch national synchronized swimming team.

Being a woman at Mines during these times must have been rather difficult. Because of their intellectual ability and their untiring perseverance, however, Caldwell, McDermut, and Davis were accepted by their male classmates as equals. In addition to earning their degrees, Caldwell, McDermut, and Davis opened the path of coeducation at Mines that at present is being trod by hundreds of women.

If Mines graduated its first women during this period, it also gained its first monetary gift from the private sector. In 1900, Winfield Stratton, then a member of the Board of Trustees, in a meeting in Colorado Springs, presented Chauvenet a check for $25,000 to benefit the school, a quite large gift for that time. Stratton, a carpenter by trade, had come to Colorado in 1868 seeking gold. On 4 July 1891, he staked what came to be known as the Independence Mine near Cripple Creek, starting the Cripple Creek gold rush and turning Stratton into a very rich ex-carpenter. The funds were eventually applied toward the construction of Stratton Hall, now the second oldest building on the campus. The cornerstone was laid in 1902, and the building was completed in 1904.

III

The years of Chauvenet's administration were, unfortunately, not free from student pranks and squabbles between students and faculty, some of which resulted in student suspensions and even a student strike. Once, Chauvenet arrived in his office to find a cow ensconced there. Another time, Arthur

The Freshman Class of 1894. (PIONEER MUSEUM.)

Early physics laboratory.

Lakes' lecture platform was piled high with boxes and furniture. Chauvenet's demand that the guilty students apologize or face dismissal resulted in several members of the class leaving school. On 5 April 1897, another classroom disturbance involving freshmen resulted in a punishment calling for the class to report for an examination on 16 April (Arbor Day). Instead of accepting this punishment, the class turned out to plant trees on the campus. While these events may seem trivial, the Board of Trustees overruled Chauvenet and the faculty regarding the Arbor Day punishment. Chauvenet reportedly submitted his resignation but was persuaded by the board to withdraw it. The board did at this time pass a resolution that placed all matters of discipline and scholarship in the hands of the faculty and made their decisions final. While such a resolution looked good on paper, unfortunately it was not always to be followed.

During the final years of Chauvenet's tenure as president, Mines underwent yet another outburst of student unrest, again involving the freshman class. The flash point this time was the feeling of the freshmen that they were being graded unfairly. The evidence presented was the high number of examination failures in chemistry and descriptive geometry. The imbalance between the number of students admitted to Mines and the number who actually graduated was no doubt a key underlying factor in the whole situation. Between 1883 and 1902, Mines graduated an average of fourteen students per year, with a high of thirty-seven in 1901 and a low of one in 1890. An article in the *Engineering and Mining Journal* of 16 May 1903 referred to anonymous complaints that at Mines "it has long been the policy of the faculty to fill up the school with freshmen, thus getting the numbers and an appearance of prosperity, then to pluck them freely until finally a handful of unusually bright students were graduated." This practice, according to the article, resulted in a more highly valued degree and a number of "plucked" students who had to continue their education elsewhere. The faculty's point of view, as well as Chauvenet's, was that because of the general lack of adequate preparation received in the high schools, it was not unusual to use the freshman and sophomore years as a measure of a student's capabilities.

According to the Mines Catalog of 1899, students could enter Mines either with a diploma from a recognized high school or through the completion of an examination based on the following subjects:

In Arithmetic, they must be ready in the use of decimals and of Ratio and Proportion.

In Algebra, the first twelve chapters of Wentworth's "Higher" Algebra, or an equivalent.

The whole of Plane Geometry.

English of High-School standard is required, but no examinations in Literature, nor in formal Grammar. Lewis' "First Book in Writing English" is recommended.

In General and Physical Geography (one paper), such questions only are put as should form part of the general knowledge of any intelligent lad.

In Zoology, Steele's "Complete" is recommended for preparation.

These requirements resulted from the tightening of entrance standards that began in 1885. Once admitted, students were held to a fairly specific and strictly enforced set of academic procedures regarding course requirements and examinations that would, in all likelihood, be a bit of a shock to the modern-day student. The goal of the basic entrance requirements, along with the requirements of the courses of study, was to maintain academic standards commensurate with the positive reputation that the school was rapidly building

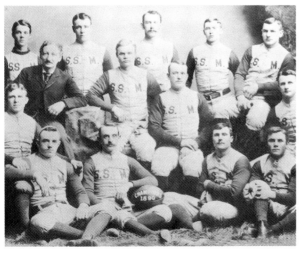

Football was a natural game for Mines students in 1890. (PIONEER MUSEUM.)

Like football, the national pastime found its way to Mines early on.

under Chauvenet's leadership. This position taken by the faculty, along with a not unusual student tendency to see themselves more or less as examples of the academic proverb that "Students must suffer at all times," gave rise to an academic conflict and an emotional one as well.

Before the smoke of this unrest cleared, a number of events occurred that might have been said to "try men's souls," namely, a strike by members of both the freshman and sophomore classes that resulted in their being dismissed from school and then readmitted; rowdyism by some members of the sophomore class that resulted in "trashing" the assay laboratory; spirited debate among students, faculty, and trustees; the failure and dismissal of six freshmen, five sophomores, and five juniors on 1 March 1902; the resignation (unaccepted) of Chauvenet; the resignation *en masse* of the faculty on 15 April 1903, along with that of President Charles S. Palmer, the man who replaced Chauvenet; and a great deal of publicity pro and contra that in sum the school would have been happy to do without. The *Denver Republican* expressed its views of the assay lab incident:

Rowdyism at the School of Mines. The Board of Regents should promptly and with emph-asis sustain the faculty of the State School of Mines in its decision suspending certain members of the sophomore class in that institution for rowdyism, insubordination, and the destruction of property belonging to the state.

The School of Mines is maintained by the people of Colorado at great expense, and students who attend it labor under the belief that by their attendance they confer a favor on the school, the faculty, or the people. (26 January 1902)

The *Gunnison Tribune* was more emphatic, urging that the students "should be promptly sent home to their papas and mammas by the faculty. The boys are altogether too fresh and a prompt dismissal would have a soothing effect upon strikes of similar nature in the future" (quoted in MM February 1981). A subsequent article in the *Republican* (undated clipping in Archive) reported that a diploma from a school "which has no superior of its kind in the world" is an emblem of that superiority and of the high standard of scholarship underlying it; and that any reduction of that standard "would impair the value of every diploma, for the public would not stop to inquire whether a man graduated before or after the reduction in the standard and requirements." Still, the fact remains that in the midst of such turmoil, it is more difficult to put the proper amount of blame in the proper places, than it is in retrospect. Once the stone of controversy is dropped into the pond, the water ripples without discrimination.

One positive, though perhaps ironic, way to view these particular incidents of student unrest, along with that of the faculty and the trustees for that matter, would be to say that Mines was becoming a part of the mainstream of American higher

Paul Meyer

In 1883, a man of small stature with a mustache, goatee, and shaggy head of hair joined the Mines faculty as professor of Mathematics. Born in Switzerland in 1854, the son of a minister, he could read Latin and Greek at age five. Graduating from Heidelberg University at age thirteen, he received a medical degree at age eighteen and a Ph.D. at age twenty from the same school. Fluent in German, French, Italian, and English, he pursued studies in mathematics, physics, astronomy, chemistry, and music (violin). Although he filled his father's pulpit on occasion, he did not have the commitment to follow a religious career. After a period of teaching astronomy, mathematics, and physics at the collegiate level, he immigrated to Denver in 1882 where he set up a medical practice.

The new Mines president Dr. Regis Chauvenet, in an effort to strengthen the school's mathematics program, persuaded Meyer to join the faculty. The two quickly became fast friends as they worked together during Mines' formative years. In addition to his teaching, Meyer carried on a part-time medical practice in Golden. In 1900, accepting the title Professor Emeritus, he formally retired from Mines though he was often seen on the campus. He continued his medical practice in Golden until his death on December 12, 1930.

Collegiate folklore is filled with stories of favorite professors who, in addition to being excellent teachers, were individuals whose personal characteristics raised them to a revered status. Paul Meyer was certainly one of these. A teacher who exhibited no arrogance in or out of the classroom, he taught his students with the same dedication that he cared for his medical patients. Perhaps the best description of Meyer is that education, a field that is notoriously fertile for controversy of one kind or another, particularly when it is recognized that the stakes are, relatively speaking, usually quite low. All that is needed for proof is a glance at almost any issue of the *Chronicle of Higher Education*, a publication that spends considerable space on controversy of one kind or another, though few, if any, controversies resemble those of Mines in its infancy. These tended to be somewhat more threatening and a great deal more dramatic.

Still, when Chauvenet left his position of president of Mines in 1902, he must have done so with the knowledge that the school, even with several incidents of student unrest, was on a strong footing and that the future looked more than promising as the twentieth century beckoned. Most of the 244 students were enrolled in the regular course program, with the special courses and special students essentially gone; the school offered the degrees of Engineer of Mines and Metallurgy and Electrical Engineering; and the faculty numbered thirteen well-educated full-time professors. Chauvenet achieved two out of his three goals while laying the groundwork for the third, that being the expansion of the curriculum to include additional basic engineering courses.

Perhaps it is unimportant what label is given to the Chauvenet era. What is important is that his presidency provided Mines with a more definite sense of direction, a stronger faculty, a more impressive physical plant, and an ever-growing national reputation as a worthy and demanding collegiate institution. As Sorgenfrei noted, Chauvenet was responsible for fashioning a course of study at Mines that turned out mining engineers

The inimitable Paul Meyer was a notable part of both Mines and Golden.

of Den Galbraith in one of the unpublished manuscripts of his "Give 'em Hell, Mines":

> Strange, enigmatic, eccentric! How do you describe a fellow like Dr. Meyer? No one understood this fellow who veiled himself in obscurity. Probably he never understood himself. What he searched for is anyone's guess. He would not have told anyone, nor would he have tried. One thing we know. He never tolerated ignorance. "Schrum! Schrum! Students are just too dumb. I can't teach them anything." Yet, he knew when he wandered down to Petries, the "Greasy Spoon" of those days, every night at 10:00 for a snack that the students would be waiting for him. He appeared not to notice that he was surrounded; students tried to think of a different approach. He knew and they knew, and they knew he knew, but it was a sort of cat-and-mouse game they played.
>
> "Schrum! Schrum!" he would snort in an air of pretended disgust. Then he would proceed to work out their calculus, mechanics, or whatever on the café tablecloth. Some nights he would stack sugar cubes on the table, trying to explain various scientific theories.

Generally referred to as "Old Chemistry," this trio of buildings was the academic center of the campus.

"who could run all aspects of a mining or milling operation. As time passed, the field became more specialized, but for the latter part of the 19th century until approximately World War I, this type of mining engineer of the old school reigned supreme." If formative best describes all of that, then so be it. If Regis Chauvenet did not turn the locomotive completely around, the strong shove he gave at least put it into motion.

Charles S. Palmer, longtime chairman of the Chemistry Department at the University of Colorado, became president in the fall of 1902. While a capable administrator, Palmer got caught in the "buzz saw" of the unrest at Mines. Palmer, unlike Hale, may have won the students but lost the faculty. As noted earlier, on 15 April 1903, the entire faculty turned in their resignations to the Board of Trustees, effective unless the president resigned. Den Galbraith, in his "Give 'em Hell, Mines" history lists the following reasons for the mass resignations of the faculty, as well as rules that they wanted put into effect:

1. A change in the rules regarding conditions and examinations.

2. The method of handling the question of intemperance.

3. Raising the standards of scholarship in faculty, requiring graduate work and practical experience in addition to college training.

4. General policy of the school in regard to relations between instructors and students.

No compromise between the faculty and Palmer on these questions could be reached; thus, in Galbraith's words, "the board dismissed the whole works, president and faculty, with the announcement that a new president would be selected and he would be allowed to select his own faculty." The *Engineering and Mining Journal* made an almost prophetic comment regarding the situation at Mines: "So the trustees are now seeking for a new president, and the latter is to select his own faculty. The gentleman on whom the honor will fall has our sympathy, for so long as the trail of the political serpent is over it all, there will be no peace for the chief of the School of Mines at Golden" (16 May 1903).

In the meantime, Horace B. Patton, professor of Geology, took over as interim president for the summer, and, in the fall, Victor C. Alderson formally assumed the presidency—no doubt with a collective sigh of relief from all concerned. Notwithstanding a resignation or two and the addition of Frank W. Trapenhagen as head of the Metallurgy Department, the faculty that Alderson chose was much the same as that which resigned in the spring. So, with a girding of loins, Mines, under Alderson's leadership, was ready to move further into the twentieth century. But the journey was not to be without a few slings and arrows, some perhaps bordering on the outrageous. Alderson, to say the least, was to become one of the more controversial figures in the history of Mines, the only president to be dismissed twice, yet to have a building named for him and to be granted an honorary degree. It is a truism that progress does not come without bouts of controversy, often along with a few bruised egos, since progressing for one person may be seen as regressing by another. Even more important, perhaps, is to remember that neither progress nor its opposite can be fully recognized until after the passage of time. This fact is not peculiar to the Mines story. On the contrary, it is a signal dimension of the history of civilization in general.

IV

Because symbols and traditions are among the most effective controlling factors of human existence, it is not surprising that American colleges and universities use them to create a defining identity and spirit of place for their students and alumni. Ridiculous or sublime, they provide a release from the tedium of study and a lifetime of nostalgia and storytelling. Through most of its history, Mines has been primarily an all-male school, enforcing traditions that reflect "maleness." Many of these traditions have been aimed at initiating a freshman into an awareness of the institution: wearing a miner's cap, wearing clothes backwards (or losing them), carrying books above the head, sounding off, and forgoing beards, mustaches, and neckties. Others required contests between freshmen and upperclassmen, usually sophomores: the freshman-

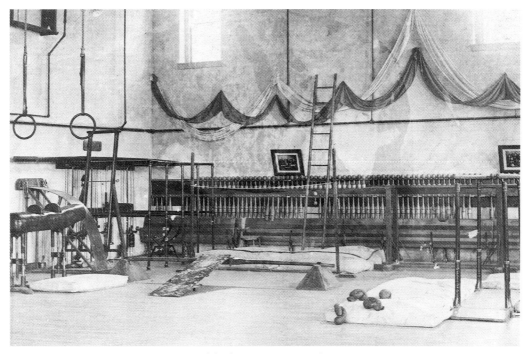

Interior of the first gymnasium on the campus.

The first woman graduate of Mines, Florence Caldwell Jones is pictured here in later years with her son.

sophomore tug-of-war over Clear Creek; the freshman-sophomore Barbeque, in which each side tried to subdue and rub dirt into the chests of the other; and the Gauntlet, in which freshmen ran between two lines of belt-swinging upperclassmen. Less combative traditions included requiring freshman attendance at football practice, painting the "M" on Mount Zion, the wearing of Stetsons by seniors, and participating in Homecoming and Engineer's Day. Athletic teams, fraternities and sororities, musical groups, drama groups, campus publications, honor societies, and a host of other organizations have all created their own long-term, multifaceted traditions that appeal to students, faculty, alumni, and friends.

Some of Mines' strongest traditions have been generated by its athletic teams, not necessarily from won-lost records but from the spirit of participation. Chauvenet himself took an ardent interest in sports and during his tenure encouraged their development. If America at the turn of the century was increasingly influenced by industry and technology, it was also becoming more concerned with the sports of baseball, football, and, eventually, basketball. Although baseball especially, but also versions of football, had been played for years, at this time each was truly finding its niche in American culture. Nowhere was this truer than on the campuses of the colleges and universities, be they large or small.

While speculation as to when the first game of a particular sport was formally played in America on any level can be an interesting pastime, it is probably not very important. As institutions of higher education multiplied across the country, athletic contests between rival schools, especially in football, did also with great rapidity. It was not long before athletics became a major factor in many institutions, to the extent that a school without organized athletics was in the eyes of many no school at all. So, where was Mines in this growth of intercollegiate athletics that not in too many years was to become a frenzy across the land?

It is important to remember that sports of one kind or another were a part of Mines from the beginning. According to Fritz Brennecke, longtime coach and director of athletics at Mines, as well as historian of

Charles S. Palmer followed Chauvenet as president and found Mines not the easiest place to be such.

Chapter 2: Beyond Silver and Gold

This group of students (1898) shows Florence Caldwell toward the left side of the second to last row.

the school's athletics, games such as cricket were no doubt played back at the time of Jarvis Hall. In "Yesterday's Heroes," one of his many comprehensive essays on the history of Mines athletics, Brennecke presented proof that this was so:

> Athletics at Mines are steeped in tradition—as much as any other part of the school's interesting history. Competition in sports began almost as soon as it was founded. In the archive of the Mines library is a lithograph of Jarvis Hall as it looked in 1868. Upon close inspection of the picture, figures may be made out, one of which is holding a cricket paddle, with others stationed at various positions on the field, including the bowler (pitcher). (MM October 1978)

Den Galbraith cited 12 November 1873 as the first public mention of athletics at Jarvis Hall when it was reported that both cricket and baseball clubs were being organized. All of these early athletic activities were quite informally organized, often drafting faculty members or town citizens to coach or manage.

Although other sports such as cricket, baseball, and rugby were played before football, the latter was probably the first sport to have regularly scheduled games and certainly the one that dominated in terms of interest. And why not? Mines was primarily an all-male school, and football is definitely a male sport with plenty of physical contact that can be measured by both sight and sound. It emulates war in many ways, complete with military-type strategies, though without the loss of life, at least purposely. Ironically, football of this period, though less complex in its strategies, was probably even closer to war. Mass play was the order of the day on fields often studded with rocks. The armor of the present-day player was nonexistent, and the lack of numbers meant that most players played both offense and defense with little, if any, rest. Three yards and a cloud of dust were more often than not a few inches and a flow of blood. While intelligent young men may have played the game, it was brawn that ruled the day. Indeed, the sport became so violent that President Theodore Roosevelt threatened to make the game illegal unless the violence was reduced and more formal rules were instituted to make it less dangerous. One might speculate what would have happened to the sport if Roosevelt had not intervened.

Football, however, dangerous or not, was coming to Mines and became over the years a vital part of the Mines spirit. The *Transcript* of 12 October 1888 noted "A game of ball will be played in Golden next Saturday by members of the Denver High School and the School of Mines boys." A week later, the results were published: "On last Saturday a very scientific game of ball was played at this place between the Denver High School and the School of Mines boys of this place. The score stood 6 to 36 in favor of the high school." Not exactly a detailed coverage nor an outcome that bode well for football at Mines; but the game was played, and because it was, athletic competition was on its way to becoming a vital part of campus life.

While the above game may have been very "scientific," it is not considered the first football game of record for Mines. That honor goes to the scoreless-tie contest between Mines and Denver University that was also played in the fall of 1888, at least according to the list of Mines football scores over the years compiled by Dave Johnston, manager of athletics at Mines from 1914 to 1952. The following year, Mines played but one game, defeating Colorado College 14 to 6 on a sandy field along Clear Creek in Golden. Although it was the only game of an abbreviated season, it began a rivalry that lasted for some ninety years, a rivalry between schools whose players participated primarily for fun and pride.

As was true in many colleges and universities at this time, most of the teams at Mines were organized on a club basis. Regaled in makeshift uniforms that would make many of today's gridiron gladiators shudder, they played against whatever competition they could find, ranging from high schools to semiprofessional teams, and to other colleges. In 1898, with George Beadle as coach, the Mines football team won a record eight games without a defeat. Because no field in Golden had adequate seats for spectators, Mines was forced to play its home games somewhere in Denver. For practice, the team used what is now called the Kafadar Commons.

In 1900, in a project that indicated an interest in more than academic endeavors and buildings, Mines joined the newly formed Rocky Mountain Intercollegiate Athletic Association (forerunner of the present-day Rocky Mountain Athletic Conference). Other members of this early version of an athletic conference were Colorado A & M, Colorado University, Denver University, and Colorado College. As membership in this conference changed several times through the ensuing years, Mines achieved the honor of being the only charter member remaining today. The Mines spirit, however, in subsequent years, was to receive much of its impetus on the playing fields and courts as its teams competed with intra-state rivals from the conference and other teams that were to become members—winning the championship in football a stunning eight times.

As might be imagined and as the history of collegiate athletics in general has shown, there was in the early years little regard for details such as eligibility rules, as Brennecke made clear:

> Some of the players were recruited from the town and were not even registered for classes, or at best, perhaps taking a short course in

Built in 1900 with a gift of funds from Winfield S. Stratton, Stratton Hall stands today as the second oldest building on the campus. (PIONEER MUSEUM)

assaying. Following the turn of the century rules were still fairly lax. Gordon "Shorty" Ellsworth, who coached four championship [football] teams ('04, '05, '06, and '07) had a fullback named Lloyd Nordenhold who'd been an All Big Ten at both Northwestern and Chicago universities before transferring to Mines, where he played for 4 more years.

Where this early version of Notre Dame's "Gipper" played after he left Mines is not evident. Be that as it may, just as Mines had grown into a reputable academic institution during Chauvenet's years as president, it was gaining regional prominence in its athletic endeavors. Before long, however, athletics and the emphasis thereon eventually became a bone of contention among faculty, students, and alumni, just as it did at a number of such early stalwarts of the sport as Harvard and Yale. Indeed, Harvard's president, Charles William Eliot, in 1903 denounced the game as a "brutal and corrupting endeavor, antithetical to Harvard's educational mission." Football, he felt, "incapacitated students for intellectual activity" (see Craig Lambert's interesting article "First and 100" in *Harvard Magazine* September/October 2003). Despite Eliot's views, a group of Harvard alumni the same year spearheaded the construction of one of the nation's earliest and most revered football venues: Harvard Stadium.

Basketball came to Mines as an official sport in 1905 and was an unqualified success with six consecutive championships, two in the earlier intercollegiate association and four in the Rocky Mountain Conference. The gymnasium, completed in 1908 at a cost of $90,000, gave Mines a step-up in athletics and in bragging rights, with the first facility in the state built specifically for basketball and other indoor sports. Featuring walnut paneling and terrazzo-tiled floors, this handsome building boasted two floors and a basement, providing space for coaches' offices, the Integral Club, training rooms, the first college swimming pool in the area (forty-two by twenty-seven feet), and a basketball

The Class of 1902.

court. The oldest gymnasium in the state and the best equipped for a long time, this building attained a storied history in Mines athletics. When in 1960 it became the Mines College Union, athletic equipment and facilities gave way to overstuffed chairs, pool tables, a faculty lounge, a more attractive room for the Integral Club, and a ballroom where the basketball court had been. A structure that to this day would be worthy of the Mines campus, it was finally razed in the mid-seventies to make way for an addition to the Arthur Lakes Library.

To say that Mines was ready to handle easily any challenge that the new century might bring its way would probably be somewhat of an overstatement. Nevertheless, here was an institution that had shed its swaddling clothes in both academics and athletics, one that could look forward with confidence.

CHAPTER 3

Progress Amidst Adversity

To live content with small means; to seek elegance rather than luxury, and refinement rather than fashion; to bear all cheerfully, do all bravely; to listen to the stars and birds, to babes and sages with open heart, to study hard, think quietly, act frankly, talk gently, await occasions, hurry never; in a word to let the Spiritual, unbidden and unconscious, grow up through the common; this is to be My Symphony.

—William Ellery Channing

The above words from the New England transcendental philosopher and writer were quoted in the conclusion of a commencement address given by Victor C. Alderson at the South Dakota School of Mines on 2 June 1902 shortly after he assumed the presidency at Colorado School of Mines. The title of his address was "Artist or Artisan—Which?" The major thrust of Alderson's comments on this occasion was that the graduate engineer, if he were to be successful, had to be an artist rather than an artisan. He not only needed sound technical training as a basis for continuing the required education to further his professional career, he also must be a reader, a thinker, an observer, and a writer. His character and his ethical stance, moreover, had to be beyond reproach. The engineer Alderson envisioned would evolve from a competitive process maintained through elementary school, high school, college, and onto the job, until finally the artist-engineer would emerge. It obviously requires more than a commencement address to define the difference between an artist and an artisan. But few would disagree with Alderson's thesis that a true education in engineering should be a blend of high technical skill and a constant personal enrichment, resulting in a career of both substance and service. While Alderson's words were intended for the graduates of the South Dakota School of Mines, they were also most certainly aimed toward Golden, Colorado, and the students of the Colorado School of Mines.

I

An honors graduate of Harvard University in 1885, Victor Alderson embarked on a career in education, serving first as a superintendent of public schools and as a teacher of high school mathematics. In 1893, he joined the faculty of Armour Institute of Technology in Chicago as a professor of mathematics. From 1898 to 1903, he served as dean of Armour Institute, receiving from the school an honorary Doctor of Science degree in 1903. An impressive looking man with black hair and beard, Alderson resembled Ulysses S. Grant. He had strong

Opposite page: Knowledge of first aid and mine rescue were of prime importance in the preparation of mining engineers.

Mines' growth is reflected in this picture of the 1902 student body.

convictions regarding education, particularly at the collegiate level, and expected Mines graduates to be gentleman engineers, educated and not merely trained. A proponent of correct writing, Alderson regularly taught a course in English. He also expected students to dress neatly. Whether he instituted a formal dress code is questionable, but students sang a song about Alderson that ended with the following admonition:

> The boys must wear their stand-up collars
> And also their dress suits,
>   And if they are going to go to school with me
> They have got to cut out the boots.

However seriously one might interpret the words from that song, or for that matter, how seriously Alderson himself believed them, he obviously was a man of many dimensions. He projected his views with a bluntness considered overbearing by faculty members, trustees, and alumni, many of whom had their own levels of egoism. It was not long before they were all caught up in a power game that could be described as an academic version of "dodge 'em cars," in which highly educated, highly motivated, and highly dedicated individuals conflicted with each other in situations that escalated out of control and demeaned all involved.

Although he had the discretion to choose his own faculty, Alderson generally retained those of the preceding administration, probably realizing that such strategy would be the most efficient, given the circumstances under which he accepted the position of president. Having grown rather dramatically since Chauvenet's early years, the faculty numbered twenty, including Alderson, and covered a fairly broad spectrum: Chemistry, Electrical Engineering, Geology, Machine Design, Mathematics, Mechanics, Metallurgy, Mineralogy, Mining, Physics, Qualitative Analysis, and Surveying. Exclusive of Alderson's salary of $5,000 and with the absence of any for Meyer, the average yearly remuneration for the remaining eighteen members of the faculty was $1,666. In addition to English, Alderson taught a course in stocks and bonds while becoming the first Mines president to spend most of his time in an administrative capacity, another indication that

Mines, with its growth, was adopting, for better or worse, a more bureaucratic structure to fulfill its mission during the new century.

Numbering 291, the student body comprised 50 seniors, 58 juniors, 86 sophomores, and 97 freshmen, a distribution reflecting almost a 50 percent attrition rate from the freshman year to the senior year. Colorado residents led the way with a total of 215, with twenty-two other states and four foreign countries making up the remainder. Of 76 students from outside Colorado, 18 were graduates of other institutions of higher education, and 58 had more than a high school education. The actual graduating class of June 1904 numbered 30. Thus, while still mostly a Colorado institution, Mines was gaining a student body more broadly based and better prepared.

Hardly imposing, the main part of the campus consisted of four academic buildings, grouped in a two-square-block area bounded by Fourth Street (now Fourteenth) on the north, Fifth Street (now Fifteenth) on the south, Washington Avenue on the east, and Mary Street (now Illinois Street) on the west. These were the original building (Old Chemistry), completed in 1880, along with its additions of 1882 and 1890 (together called Main Hall); Engineering Hall, completed in 1894; the Assay Laboratory, completed in 1900; and Stratton Hall, completed in 1904. Other campus structures included the president's house, a boiler building, and a shop building. One early innovation of which Alderson spoke highly in his Biennial Report of 1904 was the Summer School for Miners founded that year through the cooperative efforts of Mines; the Massachusetts Institute of Technology; and Harvard, Yale, and Columbia Universities. This six-week summer school held at Silver Plume offered trips to several other mining areas and ore-treatment works.

During the first nine years of Alderson's initial term as president, essentially the first decade of the twentieth century, positive developments occurred in many areas at Mines. In terms of curricular offerings, the degree Metallurgical Engineer was added in 1904, while that of Electrical Engineer was dropped, evidently in an effort to return to the academic basics upon which the school was founded. This fact was recognized in an article by George H. Hutt titled "Colorado's Great School" that appeared in the *Rico News* (21 December 1907), a publication of the Anaconda Copper Company. Tracing the lines of Mines' development, Hutt made the following observations:

> All this variation was probably wise at the time, but one persistent force is evident throughout the entire history of the school—the use of the mines, mills, and smelters of the state as the outside, to-be-visited laboratory of the school. As a resulting consequence, the school has taken in recent years its present dominant characteristic, that of a high-grade school devoting its entire energies and its full income to mining and metallurgy, admitting only regular students who are candidates for a degree. So high grade and efficient has it become that the influx of graduates of other colleges, as well as of advanced-standing men,

Victor C. Alderson took over the Mines' reins in 1903 and was to be one of the more dynamic and controversial of Mines' presidents.

Built with money from Simon Guggenheim (1906), Guggenheim Hall remains the focal point of the campus.

has led to the organization of a graduate course in mining and metallurgical engineering. Graduates of the leading colleges and universities of the United States, as well as of the Central Technical College of London and the Imperial Technological Institute of Tokyo, Japan, are now enrolled among the students.

Although some technical educators debated the advisability of large-scale laboratories for metallurgical instruction, Hutt pointed out that Mines needed just one more thing to keep it above criticism, and that was an ore-dressing and metallurgical experimental plant of commercial size. A few years later (1910), both Mines and the Royal School of Mines in London had such a facility. Alderson detailed the physical features and research potential of the plant just completed, concluding with an interesting observation on the role of government in relation to industry and education:

> The possession of this plant enables the School to enter that broad field of usefulness that lies outside the distinct educational work of educating mining engineers. In every industry there are many problems, too large for the individual operator, which naturally can be solved best by the state or federal government. The government surveys the coast lines, maps the shoals, erects the lighthouses, prepares tables which show high and low water at all points on the coast; but, after supplying all this information, the government leaves the captain to manage his own ship. (*Metallurgical and Chemical Engineering*, May 1912)

Alderson saw the plant as a real boon to the school with its various features allowing considerable flexibility of function. It provided not only excellent laboratory facilities for students, particularly seniors writing theses, but also facilities for faculty research and for commercial ore testing. Two appropriations from the state, one of $75,000 and one of $50,000, were made available for constructing and equipping the plant. With some blips along the way, the plant served Mines and industry for over fifty years. Because of environmental problems, it was closed and eventually torn down in the 1980s.

A more significant Mines landmark, Guggenheim Hall, was completed and dedicated in 1906. Simon Guggenheim, son of financier and industrialist Meyer Guggenheim, in his successful campaign for United States senator from Colorado, made a series of donations to various Colorado public institutions, of which Mines received $80,000 to construct a much-needed facility. The result was a building that housed administrative offices, an assembly hall, the library, the geology department, and lecture rooms. Ultimately, it became the focal point of the Mines campus and remains so to this day. When Simon Guggenheim died in 1941, he left an additional $100,000 to Mines, and in 1969, his

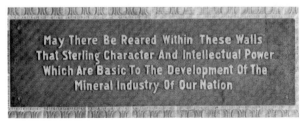

This plaque is in the atrium of Guggenheim Hall.

Chapter 3: Progress Amidst Diversity

The library was located in Guggenheim Hall.

wife gave Mines yet another $100,000 to be applied to renovation of Guggenheim Hall. Assay facilities were also enlarged during 1905 and brought the total value of buildings and equipment, as of July 1906, to $450,000, a figure that few, if any, in 1874 would have believed possible.

Two important Mines publications arrived on the scene during this period: the *Colorado School of Mines Quarterly*, a scholarly publication, in 1905 (replacing the *Bulletin*) and the *Colorado School of Mines Magazine*, an alumni association publication, in 1910. The alumni publication changed its name a few years later to the *Mines Magazine* and increased its issues from nine per year to twelve. On the academic side, in 1905, two professional fraternities established chapters on the campus: Alpha chapter of Tau Beta Pi, an engineering scholarship fraternity similar to the liberal arts Phi Beta Kappa; and Gamma chapter of Theta Tau, an engineering fraternity. Together, these publications and organizations were clear indications that Mines was making its campus worthy of its growing international reputation.

II

Alderson, as his writings often indicate, was interested in mining education as it was and should be. He was also focused on bringing the respect to mining engineering as a profession that he thought it deserved. In bringing these two interests together in a comprehensive article titled "Mining Engineering Education in the United States," Alderson attempted to explain why mining engineering was often considered the least progressive of the various fields of engineering. Following a detailed examination of the number of schools offering mining engineering programs in the United States, he counted three thousand total students enrolled in those programs. This rather low number of students studying mining engineering had, according to Alderson, four basic causes: (1) a continued lack of appreciation of the need for mining and metallurgical engineers; (2) a lack of financial support of new schools in the West where mining engineers were in demand; (3) the failure on the part of legislators to recognize that mining instruction could extend beyond mere metal to the broad field of nonmetalliferous mining; and (4) the unorganized condition of the mining industry itself. Moreover, the mining engineer, according to Alderson, needed a broad education:

> In addition to the knowledge of the electrical, mechanical, and civil engineer, he must have the training of a chemist, a metallurgist, a geologist, and a businessman; besides, he must have, as it were, a stomach like that of an ostrich, a heart as vigorous as a force pump, and a physique like Sandow; in the camp he must rough it with

Guggenheim Hall provided Mines with its first real auditorium.

> ## After the Boulder Game
> ### (1914)
>
> Oh! they said we couldn't do it,
> But we did—
> Yes, we did.
>
> Every Miner played like—well,
> Not a single hero fell.
>
> Oh! they said we couldn't do it,
> But we did—
> Yes, we did.
>
> Now they know that we can do it,
> For we did—
> Yes, we did.
>
> On her knees proud Boulder fell,
> While the Miners gave 'em—well,
> We're the Rocky Mountain Champs.
> So we did—
> Yes, we did.
>
> *Lillian Van B.*
> (Mines 6 – Boulder 2)

the Slav and the Pole, but in New York he must have the polish of a Chesterfield; besides all these qualifications, he must be a skilled manager, a practical economist, and above all, a dividend producer. (*Colorado School of Mines Quarterly*, April 1908)

Alderson certainly must have known that it would take a super engineer to meet all the above qualifications. His purpose was stimulation of interest in, and support of, mining education in both the practical and the ideal sense.

Two short courses were introduced at this time, one for practical coal miners and one for prospectors, along with an institute for highway engineers and road builders, all of which reflected Mines' willingness to address the needs of the surrounding community. Another curricular-related feature that was to take on an added significance a few years later was the creation of the Engineers Company of the National Guard of Colorado that, with Alderson's blessing of academic credit, offered training in drill and military tactics to Mines students, who would become the nucleus of this unit. Students, moreover, would soon be exposed to another kind of training as part of their Mines education: physical training. Alderson expressed the following justification of this addition to the curriculum: "In order to more completely meet the demand for whole-souled men, so essential to engineering usefulness, a department of physical training will be established shortly. The course will be systematic and progressive, intended to remedy common physical defects and to foster a condition of vigorous health and development of correct habits of carriage and action" (1912 Biennial Report).

While the addition of the Experimental Plant served to bring Mines closer to the practical problems of mining, Alderson still felt the need for a curricular revision that would strengthen further the preparation of mining engineers. He recognized that the field of mining had increased both in the number of metals mined and in the quantity of coal being mined. Milling, metallurgy, and mining

The Gymnasium, built in 1908, provided Mines with a state-of-the-art building for physical education and basketball.

Chapter 3: Progress Amidst Diversity

The I-Club had rooms in the Gymnasium.

geology were rising in importance. All these factors demanded an expansion of Mines' facilities and curriculum. Implemented in 1911, his new plan emphasized basic courses in chemistry, physics, mathematics, machine design, geology, and surveying to be taken during the freshman and sophomore years. In the junior year, students would select one of the following courses of study: Metals Mining, Coal Mining, Metallurgy, or Mining Engineering, upon the completion of which a single degree, Mining Engineer (M.E.), would be granted.

While the above was ostensibly a four-year program, many students required five years to complete it. Alderson felt that such an extension of study would not be a negative factor but would indeed produce stronger mining engineers. A summer session was made available for those falling behind or for advance-standing students who might enter with some deficiencies. Although part of the faculty and alumni feared a differentiation of courses and especially degrees and were opposed to a five-year program, Alderson believed his new curriculum was a step forward. Far from lowering the standard of the school, he felt that, combined with the Experimental Plant, the new curriculum significantly increased opportunities for a strong mining engineering education. It is not difficult to see additional seeds of dissension beginning to sprout between Alderson and members of the faculty on the one hand and a group of concerned alumni on the other.

Undaunted and Undefeated

The Mines football team of 1914 came from a school with the lowest enrollment in the Rocky Mountain Conference and with the most demanding course of study. Moreover, the team was the lightest in the conference. Given these statistics, few dreamed the team would go undefeated and win the conference championship. But it did, compiling the following record under Coach W. J. Hanly:

Mines	25	University of Wyoming	0
Mines	14	University of Utah	6
Mines	18	University of Denver	0
Mines	6	University of Colorado	2
Mines	19	Colorado A & M	0
Mines	7	Colorado College	7

To complicate matters, the team played its last three games in twelve days. To balance the year's athletic efforts, the Mines basketball team compiled a record of zero wins and eight losses.

The swimming pool was in the basement of the Gymnasium.

During the 1911 academic year, simmering faculty unrest began to boil. Faculty complaints to the Board of Trustees resulted in a marathon meeting of the board and the faculty on 25 May that ran uninterrupted for ten hours. At this meeting, the trustees questioned each faculty member and gave him the opportunity to explain his discontent. At the end of the meeting, however, Alderson was still president although three faculty members had submitted their resignations: A. J. Hoskins, head of the Mining Department; A. C. Smith, associate professor of mining; and C. E. Smith, associate professor of geology. Since no record of this meeting is available, it is not clear what specific complaints the faculty presented.

While the outcome of the meeting was dramatic, it did not reach the core of the problem; and in less than two years, faculty unrest again flared up, this time stoked by differences of opinion between the faculty and alumni on one side and other members of the faculty and Alderson on the other regarding the curriculum in general, as well as methods of teaching individual courses. Another question that proved divisive was the place of athletics on the campus. Some members of the faculty who had been consistently opposed to the favoritism that athletics were receiving at Mines were instrumental in getting one of their number appointed to the school's Athletic Board. Adding to the tempest was the definite clash of personalities among Alderson, trustees, faculty, and alumni and the ultimate question as to whether Alderson was qualified to fill the position of president. Shakespeare's three witches in *Macbeth* could hardly have stirred up a more destructive brew.

In their May 1913 meeting, the Board of Trustees evidently decided that Alderson was indeed not qualified to be president and refused to reappoint him. In addition, they dismissed three faculty members: George W. Schneider, an alumnus and head of the Mining Department; Carl A. Allen, an alumnus and associate professor of geology; and G. Montague Butler, an alumnus and associate professor of geology. It appears these three men were dismissed for no reason other than the feeling that they would not be a positive influence in the anticipated healing process. Given the resignations of 1911 and these three dismissals, the areas of mining and geology had taken quite a hit in terms of faculty in just two years. In a final motion, the board gave Regis Chauvenet, who had returned to Mines as a special lecturer, the title of president emeritus and faculty member William G. Haldane, the title of acting president, with the optimism that these two men, working in a dual capacity, could control the controversy causing the school negative publicity both locally and nationally. Haldane received his bachelor's degree from the Case School of Applied Science. He also completed work at Columbia University and the University of Denver

The basketball court in the Gymnasium was also used for physical education classes.

One of Alderson's favorite accomplishments was the building of the Experimental Plant that for many years played an important role in mineral and metallurgical research.

prior to his arrival at Mines in 1900 as an instructor of mathematics.

To ease the transition, Chauvenet and Haldane were given carte blanche to choose a faculty and to revise the curriculum. Chauvenet declined to participate in reorganizing the faculty, noting that four of its members were men he had appointed; four were quite good friends of his; and four he did not know at all. Be that as it may, the atmosphere at the alumni banquet the evening of 23 May 1913 was one of relief. The theme that several speakers expounded was that of dismissing the disruptive differences of the past and working for the good of the school and all concerned. It was a theme that soon would be forgotten.

Whatever the theme celebrated by the alumni, Alderson, for his part, was not ready to depart quietly. Indeed, not only did he refuse to resign, he also refused to accept the letter of commendation the trustees had prepared for him. From his standpoint, the entire effort to remove him from his position resulted from a political conspiracy between a local trustee of the school and a clique at the General Assembly. He said the fight, as he called it, started about four years previously and inten-sified as time passed. Although no record of the document can be found, Alderson argued that 190 students out of the total enrollment of 226 signed a petition one week before his dismissal to keep him in office. Frederick Steinhauer, president of the Board of Trustees, commented that the decision to terminate Alderson was based strictly upon what the board felt was best for the school and that no charges were made against him. He also expressed his personal high regard for Alderson.

III

Although this controversy consumed a great deal of emotional energy among many people concerned with Mines directly and indirectly, the school, thankfully, was progressing in its day-to-day existence as a collegiate institution. Freshmen entered with some trepidation, sophomores eagerly awaited their chances to intimidate the freshmen, juniors went about with their self-satisfied demeanor, and seniors looked forward to graduation (if not sooner, later). Classes were taught, the Glee Club's call for candidates garnered twenty-five singers, traditions were developed and followed, and Mines men did battle on the fields

A Year to Remember in Sports

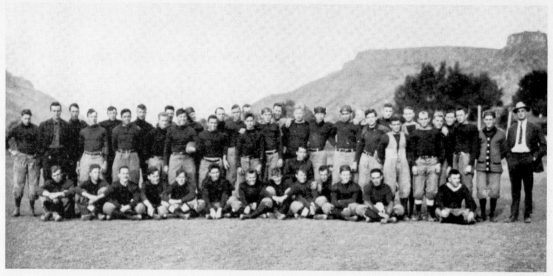

The 1912 football team won the Rocky Mountain Conference championship.

The 1912–13 year was one of the better years for Mines athletic teams. The football team was especially strong and posted eight victories against one loss, a record good enough to win the Rocky Mountain Conference championship. The lone loss came at the hands of the University of Utah early in the season. Near the end of the season, Mines and Utah were tied for first place in the conference. Mines, on its way to California to play Pomona College, stopped over in Salt Lake City to see the Thanksgiving match between the University of Utah and Utah Agricultural College. The game ended in a tie, giving Mines the conference championship. Though not playing at their usual high level, Mines defeated Pomona 13 to 0. They closed out the season with a 34 to 7 rout over another Pacific coast team, Whittier College.

The baseball team completed a season of six victories, including a forfeit win over Colorado Agricultural College when the latter used an ineligible pitcher. The final game of the season against the University of Colorado was a nail-biter. At the end of the ninth inning, the game was tied at 6. After two scoreless innings, a home run by Johnny Davis in the first of the twelfth with one man on base put the Miners ahead. Colorado answered with a single run in their half of the inning, giving Mines an 8 to 7 victory and the conference championship.

The basketball team had a bookend season, losing both the first and last games. In between, however, they easily put away five opponents and edged the University of Colorado 21 to 20, gaining revenge for a loss in the season's opener. Perhaps overconfident that they could clinch the conference championship in the final game, the Miners fell to Colorado Agricultural College 39 to 21. Together, the football, baseball, and basketball teams chalked up twenty wins to three losses. Truly a year to remember.

and in the courts of sport. In short, the Mines spirit was not to be denied by bureaucratic infighting.

Den Galbraith in his article "Give 'em Hell, Mines" traced at least some of the trouble over athletics to the days when "Shorty" Ellsworth cut a few corners regarding the academic background of some of his football recruits. "It can be noted," said Galbraith, "that many of the athletes during Alderson's 10-year stint never graduated from school." Again, it was common for colleges and universities of this period to be lax in matters of athletic eligibility. It was also a period when students and alumni were very much caught up in the frenzy of intercollegiate athletics, to the point where few of them were concerned about such minor factors as the eligibility of their Saturday afternoon heroes. Mines students and alumni were no different. And why should they have been? Mines, at least for a time, was on a roll.

"Shorty" Ellsworth, who stood well over six feet and had played football at the University of Chicago, was hired as football coach in 1904. He proceeded to win a string of four Rocky Mountain Conference championships, putting Mines on the gridiron map. A team's schedule in those days was obviously not the eleven or twelve games now played in a single season, so at first glance Ellsworth's record of fourteen wins, one loss, and four ties over that four-year reign may not seem so impressive. But in the next seven years, Mines boasted three more championship teams: 1912 (William Johnston, coach), 1914 (William Hanley, coach), and 1918 (Irving Barron, coach), with a cumulative record of fifteen wins, one loss, and one

The 1909 football team lines up for practice.

Freshmen were required to attend football practice sessions to cheer the team on.

tie. Adding Ellsworth's record brings the number of championships to eight, with a total of twenty-nine wins, two losses, and five ties (a winning percentage of .935, excluding the ties), certainly a record to make Mines a team to reckon with and to fear.

Mines football teams over the years have played in many contests that have been considered "games to remember," games that went right down to the wire before the outcome was decided or, on the other hand, were such routs that the scoreboard could not keep up. The game between Mines and Denver University on Thanksgiving Day 1913 at Union Park was certainly not a game of which the outcome was in doubt to the last seconds. In fact, Mines defeated Denver 49 to 7 in a veritable rout. An interesting aspect of this contest, at least in a creative sense, occurred after the game in an unidentified newspaper report by one Eugene Fowler. It is not clear whose side Fowler was on, but his story of the game has to rank with some of the best sports prose ever written, including the later piece by Grantland Rice in which he "christened" the Notre Dame backfield the "four horsemen" (18 October 1924). Though a bit long, Fowler's piece deserves full play:

> A sky of lead-gray with dark cloud banks dragging across the sun, 2,500 rooters, and a Marathon race in which eleven boys in blue

The Glee Club provided the campus with a touch of melody.

triumphed, were features of the last act of Denver's 1913 gridiron drama. The Colorado School of Mines swamped Denver University with a score of 49 to 7 at Union Park in the Thanksgiving Day clash.

It was a parade of advancing blue against a retreat of crimson and gold from whistle to whistle. Denver University had about as much chance of stopping the charge of the Ore Diggers as a polar explorer would have in lighting an asbestos bonfire with an icicle.

Let no man say the Denver boys lacked in gameness. They struggled and struggled well. They were alert every moment as was evinced in the last minute of the parade when Pete Miller of the University burst forward from his interference of crimson beef and shot down the field for a touchdown, Denver's only scoring play.

It was then that the Denver stands sent forth a roar of joy. Up until that last minute Denver collegians had crouched under the bite of the November air, pleading with their team to hold the scrappy Miners. If their team had displayed any football foibles in the greater part of the game, that last minute and that touchdown redeemed the warriors of the Crimson and Gold.

That the best laid plans of weather prophets and gridiron dopesters "aften gang agley" was shown by the overwhelming defeat administered by the metal seekers to the Sky Pilots. Some experts had said that Denver University would give the Miners a run for the money. They did give them a run, but it was in the wrong direction.

Some had said that the "weakened" Miners, crippled from the punishment awarded them in the Boulder game, would enhance the chances of the Ministers. But no one was seen on crutches in the Mines' lineup.

"Are we crippled? Hell, no," shouted the Mines constituents. A rather blunt slogan for delicate ears, but the earth delvers felt that way and no national board of censorship was nigh to prune their yells. And the Mines yells are not full of prunes. (Reprinted in MM November 1913)

Very much alive at this time was the competitive zeal displayed by student backers of their teams in the Rocky Mountain Conference. Whatever could be done to leave a "Mark of Zorro" on an opponent's premises or to steal something from those premises was surely fair game. Actually, it was a throwback to scoring "coups" against one's enemy. So these students, who saw themselves as justified extensions of their teams, pursued the foe in their own fashion. If they were not able to "slug it out" in the upcoming contest that would be decided by points scored on the field or court, they would determine to embarrass or anger the opposing school by painting signs on its buildings or stealing mascots. All in fun? Yes, to a degree, but often there was a fine line between all in fun and at least part in anger. Many are the stories told among Mines alumni regarding exploits directed against an opposing school, more often than not in Mines' case the university in Boulder or the one in Denver.

Usually during the week preceding a big game, the home territory and its important buildings and symbols, as well as any live mascots, were put under student guard. The "M" overlooking the Mines campus was naturally a target for opposing schools. One raid on the Mines campus occurred early on the morning of 15 November 1913 when a group of University of Colorado students broke into the Mines gymnasium to recapture a goat that Mines students had stolen at the end of the previous year's football game in Boulder. At that time, the goat was alive and was brought to Golden. After he grazed on the campus for a couple of days, he was killed and stuffed, becoming in death Sir Billy, a prize trophy. Discovered by a night watchman, the Boulder group fled without the goat. Less than a week later, on 20 November, three days before Mines was to meet the University of Colorado on the gridiron, another group from Boulder was caught painting a U of C sign on the back of Guggenheim Hall. Most of the group fled in a car over the new road up Lookout Mountain, but two were captured. They were kept in a fraternity house until morning when their heads were shaved, Mines was painted on their foreheads, and they were dressed in painter's overalls. After being taken into Denver and photographed by a newspaper cameraman, they were put on the train to Boulder. That afternoon, a mass meeting of Mines students sought better plans for guarding against any attempted revenge efforts from Boulder. Mines lost the football game 20 to 0.

IV

Following the year of the dual presidency of Chauvenet and Haldane, the Board of Trustees appointed Dr. William Battle Phillips president. A graduate of the University of North Carolina, Phillips also studied at Freiburg, Germany, in the late 1880s. He taught mineralogy at North Carolina and chemistry at the University of Alabama, after which he worked in industry until coming to Mines. One of the more prolific publishers among Mines presidents, he had almost three hundred articles in

These students from China found a collegiate home at Mines (1913).

William George Haldane succeeded Alderson as Mines president.

various scientific journals to his credit. Phillips, in a few remarks of greeting to the Mines alumni, summarized the challenges of leading Mines:

> No sane man believes that he or anyone else can conduct important affairs without making mistakes. But we have a right to expect that he shall profit by his mistakes and not make the same one twice. He must avoid the habit of making mistakes. His batting average, so to speak, cannot be 1000, but it must be at least 0.510.
>
> At the beginning of this new administration, we bespeak the sympathetic and active interest of the Alumni. The school lives in them, it derives its reputation from them, it is judged by them. What the Alumni are the school is. If they are indifferent towards its needs and towards its daily life, it will languish. If they are not eager to maintain high standards of scholarship and of service, these will inevitably suffer. If they do not love and cherish their Alma Mater, who shall stand by her side? (MM August 1915)

Whatever aspirations Phillips, the trustees, the faculty, and the alumni held for the new administration were short-lived. His "batting average" aside, Phillips lasted just two years in the pressure cooker that Mines had become. Spirited Mines students presented another side of collegiate life for which he was hardly prepared. During 1915 and 1916, it was not unusual for students to ride around Golden in cars while shooting off revolvers and throwing lighted sticks of dynamite onto the streets, causing Denver newspapers to refer to Mines' athletic teams in that era as "Dynamiters" and "Blasters" rather than Orediggers. For whatever reasons, Phillips did not desire to "go to battle" with Mines students for a third year. Following his resignation, the Board of Trustees appointed Howard C. Parmalee president. Parmalee, a graduate chemist from the University of Nebraska, who had worked in industry and had edited *Mining Reporter* and *Western Chemist and Metallurgist*, must have arrived at Mines with a sense of trepidation, given the flying sticks of dynamite and

An announcement of Senior Day—1913.

the plethora of dismissals and resignations that had occurred in the recent past. And well he should have.

The celebration of spring in one fashion or another is commonplace among American colleges and universities though perhaps few could have rivaled Senior Day at Mines during the period in question. The tradition of Senior Day began on 18 March 1912 and was designed to bring students and faculty together in some good-natured tomfoolery, a break from the rigors of study. Although the exact day was secret, it generally fell near St. Patrick's Day, perhaps because St. Patrick was thought to be the patron saint of miners. On the morning of the chosen day, a blast of dynamite announced the beginning of festivities, which featured a parade of seniors dressed in humorously outlandish costumes and usually ended with a baseball game between students and faculty. Special senior detectives served warrants on chosen faculty members to appear in a mock court. In the first Senior Day, President Alderson was found guilty of carrying

Senior Day 1913–defendants awaiting verdicts. President Alderson in the light suit.

concealed syllogisms and was fined cigars all around. Needless to say, alcohol flowed freely.

Senior Day in 1917 was celebrated on 23 April with drinking, starting fairly early in the morning. By afternoon, many students were feeling little or no pain. During the baseball game, the sophomores began a favorite campus sport called "wrinkling,"

The "M"

While there has been disagreement on the exact date that the "M" was constructed on Mount Zion, it was probably about 15 May 1908. Impressed by the "U" of the University of Utah, some Mines seniors began a campaign to put the Mines initial on Mount Zion. In early May, Herbert A. Everest '08 and Walter R. Brown '10 surveyed the area and laid out the design of a letter 104 by 107 feet on a twenty-three-degree slope with lines ten feet wide. Their goal was an "M" that would be symmetrical from every angle of vision. At 7:30 in the morning of the appointed day, a large number of students led burros loaded with tools and supplies for the filling of the outline with whitewashed rocks up Mount Zion. By late afternoon, the "M" was visible in all its glory from Golden below. The lighting of the "M" with railroad flares at homecoming by members of Blue Key during the 1920s led to the installation of electric lights, lit for the first time on 19 March 1932.

Because of the attempts of athletic opponents to destroy or deface this symbol of Mines, careful guarding by members of Blue Key and other student organizations has occasionally been necessary. When caught by "posses" of Mines students, such culprits were ceremoniously given shaved heads with a silver-nitrate "M" painted thereon—pretty much all in good fun. Students from both the University of Colorado and the University of Denver had this experience impressed on their heads and their memories.

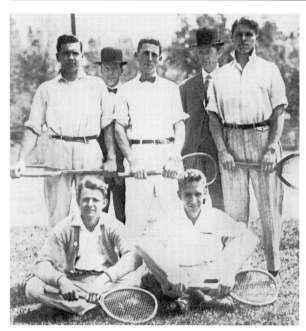

Tennis was a minor but popular sport in the early 1900s.

an activity involving the forceful rubbing of dirt and stones into the bare abdomens of prostrate freshmen. On this particular day, however, a faculty member was "wrinkled"—an indiscretion that later resulted in five sophomores being suspended. After numerous meetings that included students, faculty, the Board of Trustees, some residents of Golden, and even the governor of Colorado, the freshman, sophomore, and junior classes all dropped out of school. Although the seniors wanted to join the exodus, they were so close to graduation that the underclassmen persuaded them to remain in school. On 4 May, the faculty passed a resolution permitting all students who returned to school by the opening of the fall term to receive full credit for work already completed. This resolution apparently included the five who were dismissed for "wrinkling." (See MM May 1917 for an extended article on the above events.)

Parmalee, in an addendum to the MM article, spoke out sharply about the activities of the students on Senior Day and the lack of sufficient punishment exacted of them. He was concerned particularly that the faculty had caved in, allowing all students, including the five formally dismissed, to re-enter school in the fall term. He also thought the excessive number of people involved in ironing out the incident only complicated matters and closed with the following:

> Every selfish personal interest would dictate that we yield to the [student] demands; but I submit we are not confronted with a question of personal expediency, but of principle. There should be no doubt as to whether the authorities or the students are to run the school. If the authorities are incompetent, they should be removed, but they should not be overridden.

Parmalee could not have been more prescient. The students argued that the "wrinkling" of the faculty member was part of the general "wrinkling" among the students and was certainly not intended to be a premeditated personal assault but merely a "boyish prank." They felt that, though punishment of some kind was deserved, the particular form imposed more than fit the crime. As a result, the entire student body voted to withdraw, feeling that the unrest and dissatisfaction that had existed since the beginning of the year would continue, and that it would be a waste of time to stay in school.

Parmalee became part of the string of dismissals and resignations that must have looked to an outsider

The Mines Athletic Board of the period.

Senior Day scenes—1914.

like a weird version of musical chairs. Certainly, there were outsiders who noticed. An editorial in the *Engineering and Mining Journal* made no bones about its view of the events at Mines:

> The striking students behaved in a disgraceful way. No less disgraceful was the behavior of the trustees. Indeed, theirs was the more disgraceful for they were older men and should have known better how to act. The president of the institution should have been supported even if he had been an incompetent. In this case, he was not only a highly competent and professionally esteemed gentleman and engineer, but also his position in the school controversy was correct in all respects.
>
> Ex-President Parmalee knows that his unhappy experience at Golden did not cause him to depreciate a whit in the esteem of his professional colleagues. The persons who lost were the incompetent majority of the Board of Trustees and the unscholarly, striking students who will, like the mutinous soldiers of Russia, individually wear a black band of disgrace until they have redeemed themselves by some signal exhibition of correct conduct. But for redemption of the Golden school it will be necessary for the governor of the state to appoint a new board of trustees, and the new board must use strong disinfectants to remove all odors of politics and petty interests. (18 August 1917)

The editorial also castigated the system of annual selection of president and faculty that was standard procedure from the beginning of the school. An editorial in the *Mining American* also aimed at the Mines Board of Trustees:

The Great Barbecue
From the November 1908 "Retort"

How this affair got the name Barbecue is not known, but it was a contest between the sophomores and the freshmen.

It seems that the Freshies were especially anxious to meet their opponents, who were greatly their inferiors in numbers but not in prowess and generalship. About 7:00, they congregated at the tramway depot and, freshman-like, were making a big hullabaloo there and yelling themselves hoarse. About 7:30, they had all congregated and then marched up the south side of the campus in front of Stratton Hall about 150 strong, yelling like demons all the way and wearing themselves out with a superabundance of enthusiasm. But where were the Sophs who had so valiantly boasted the night before of their intention to tie the Freshies with rope that had a wire core? Where were those bloodthirsty bandits who threatened to use handcuffs on the poor, helpless freshmen, those brave knights who had gone around in the quiet of the night plastering the town with posters, green (which means Freshies), flaring posters which flaunted warning in blood-red ink?

A few minutes before 8:00, the Sophs came sauntering over and took their stand on the north side of the lawn. Then the signal was given and the scramble was on. At first, neither side made much headway, but after half an hour very few freshmen were to be seen running around loose on the campus. It was a fierce battle and many a sophomore was hindered while attempting to tie his man by the interference of some of the spectators. And, occasionally, he would find his rope cut after he had securely tied his opponent's hands. Occasionally, a few freshmen would break from the camp, but after one and one-half hours of hard work, every freshman on campus was tied and in the camp of the victorious Sophs.

To our way of thinking, the school needs readjustment higher up, in the Board of Trustees. No person who can show any selfish interest in the management or policy of the school should be eligible to membership on the board. Further, the board should be thoroughly removed from politics, and it should not include any man who, because of physical or commercial reasons, cannot attend all meetings and at all times give unprejudiced aid to the upbuilding and support of this state institution. This school is not the property of a village; it belongs to Colorado. (14 July 1917)

The *Mining and Scientific Press* added its view of the students who caused the ruckus on Senior Day: "What they need is military discipline. Disloyal to their alma mater, which they have brought into ridicule, mutinous to their teachers, reckless of their careers, how can they expect to become fit to direct the work of other men in mines and mills if they themselves exhibit so little self-control?" (16 June 1917).

For Mines to be highlighted in such negative ways in such prestigious journals was, to say the least, disconcerting to everyone related to the school. Nor was this spate of bad publicity to be the end of the matter. Meanwhile, however, the Board of Trustees, after conducting a presidential search that was for various reasons unproductive, found themselves once again at square one. Understandably, many potential candidates may have had misgivings regarding a position so unforgiving of its occupants. Thus, the board turned once again to Victor Alderson as the man to bring some modicum of stability to the school after a tumultuous several years. One can speculate why Alderson himself agreed to return to Mines. Did he simply enjoy his first presidency? Was he trying to defeat those he felt sabotaged his earlier career at Mines? Did he want to implement his ideas of what a technical education should be and how Mines might reflect that? Perhaps the answer is all of the above. But getting him once again on the "good ship" Mines was not exactly a smooth operation.

When the Board of Trustees, with the approval of the governor, offered the presidency to Alderson in September 1917, it is not an exaggeration to say that "all hell broke loose." His former opponents were still circling about, and they immediately

Members of the "M" Club.

View of the campus, circa 1915.

zeroed in for another kill. On 10 September, a group of alumni, ignoring their association's resolution of 1913 "to forget all past differences and that all stand together to aid and support the best interests of the school and the administration of the school at all times" (MM October 1917), succeeded in obtaining a temporary injunction from the District Court to prevent the board from ratifying its decision to hire Alderson. They argued that returning Alderson to the presidency would destroy the school. The *Golden Globe* asked "how a court could restrain a board of trustees from doing that which the law directed and required its members to do" (27 October 1917). The question became academic when the District Court refused a permanent injunction, and the Supreme Court refused to hear a further complaint. Thus, Alderson was, so to speak, back in the saddle. And so was the conflict between members of the alumni and the Board of Trustees.

"It is restful to the public mind," wrote John Vivian (later governor of Colorado) in the *Golden Globe* of 3 November 1917, "to have the atmosphere cleared and to see smooth sailing ahead. Most of those who opposed Dr. Alderson did so with honest motives, but there was malice, apparently, in the hearts of others." Indeed, Mines did sail on, even if not so restfully as Vivian no doubt desired.

Chapter 3: Progress Amidst Diversity

Dr. William Battle Phillips replaced Haldane during the "musical chair" period of presidents.

The 1906 basketball team won the Rocky Mountain Conference championship.

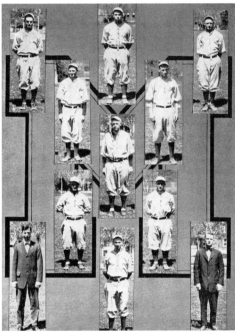

The 1913 baseball team won the Rocky Mountain Conference championship.

Dr. Howard C. Parmalee, seventh president of Mines.

National Guard unit.

CHAPTER 4

Steadying the Course

A University is, according to the usual designation, an Alma Mater, knowing her children one by one, not a foundry, or a mint, or a treadmill.
—Henry Newman

As Mines opened its fall semester of 1917, the United States had already been at war with the Central Powers of Europe since the previous April though many Americans had yet to realize the full significance. While not strangers to war, they certainly had no idea of the magnitude that modern war, total war, had reached. Nor were they conscious of exactly what this war was all about and would not be until some time after the conflict, if even then. But war is war, and the American ingenuity and pragmatism that had built the nation was put to work to help win one some three thousand miles away. When the last guns were fired and the last casualties were counted, the United States was on its way to becoming the most powerful nation the world had yet seen. It might take another war to prove it, but for certain it was coming.

Colleges and universities felt the impact of the war as much and as quickly as any other segment of American society, for many of their current and prospective students were to be enlisted or drafted into the armed forces. Many, moreover, were to become casualties. Although declining enrollments delayed plans for growth and development, most colleges and universities, even with the necessary changes in their normal procedures, carried on with their educational missions.

I

Perhaps the most significant single wartime event to touch Mines, as well as over five hundred other collegiate institutions, occurred on 1 October 1918 when male students were "drafted" into units of the Student Army Training Corps (SATC). For Mines, this meant virtually the entire student body. To the school's credit, 162 students passed the required physical exam and were organized into one company of four platoons. Since Mines, at the time, had no dormitories, the student soldiers were housed in fraternity houses and other available residences around Golden. Under the direction of a captain and two master sergeants, the daily schedule

Opposite page: Company A (ROTC) at attention.

The ROTC was an important part of the Mines curriculum.

The student soldiers were issued uniforms and were marched from place to place throughout the day under close supervision. Obviously, there was little time or opportunity for "pranks" or any activity other than that described above. While some persons may view Mines today as an academic "boot camp," it surely was such in 1918.

A complicating factor during the year was the terrible outbreak of Spanish flu that affected the entire country, as well as the world, and caused many quarantines and class cancellations. In a most informative taped interview available in the Archive, alumnus Albert Smith '22 talks not only of the military aspects of Mines at the time, but also of the constant concern of the spread of the flu. Although the flu struck Golden hard enough, Mines students at first seemed immune from the disease. But, as Smith reports, some students went into Denver for dates and returned with the flu, resulting in a quarantine of the entire school. To prevent

was rigorous because these student soldiers were required to combine military and academic work:

First Call	6:00 a.m.
Reveille	6:15 a.m.
Breakfast	6:30 a.m.
Military Training	7:00 a.m. to 8:00 a.m.
Academic Work	8:00 a.m. to 12:30 p.m.
Lunch	12:30 p.m. to 1:30 p.m.
Military Training	1:30 p.m. to 2:30 p.m.
Academic Work	2:30 p.m. to 5:10 p.m.
Dinner	5:30 p.m. to 6:00 p.m.
Athletics	6:00 p.m. to 7:30 p.m.
Supervised Study	7:30 p.m. to 9:30 p.m.
Quarters	9:45 p.m.
Taps	10:00 p.m.

Fraternities played an early and important role on the campus.

A campus view from Castle Rock.

deaths among students, the Army sent a corpsman and nurses to care for them. A makeshift hospital was set up with intensive-care space for about a dozen of those most seriously ill. Smith notes that liquor was used to assist the curing process. He himself managed somehow to dodge the "flu bug." Smith recalls the ringing of bells on the day of the armistice, but no students were allowed to join in the town's celebration. Smith earned his degree (184 credits) in Metallurgy and returned in 1947 to teach at Mines until 1966.

As noted earlier, military training at Mines dates from about 1873 when students from Jarvis Hall, Matthews Hall, and the School of Mines received instruction from George West, who had been a Union captain in the Civil War. Moreover, in 1909, a National Guard unit was formed in Golden and used the Armory as headquarters. Because many of its members were Mines students, Alderson agreed to grant academic credit to those students who successfully completed a course in military training as members of the Guard unit. With America's entrance into World War I, this unit, along with one from Boulder and one from Colorado Springs, became the First Battalion, Colorado Engineers. After a period of training at their camp in Golden City Park, the battalion was drafted into federal service in August 1917 and was soon expanded into the 115th Regiment, U.S. Engineers. After a year training at Camp Kearney, California, the new regiment set sail for France, where it joined the Sixth Corps and saw action in the Meuse-Argonne and Toul sectors, repairing and rebuilding bridges under frequent shell fire. After the armistice, the regiment was sent to Coblenz, Germany, for occupation duty. Fortunately, the 115th suffered no fatal casualties. The colors of the regiment, twice decorated in France, were presented to Mines in May 1931 by an organization of veterans of the 115th. They now hang in the school's Military Science Department.

The SATC was an expediency growing from the wartime need to produce officers for the rapidly expanding military and was not intended to be a long-term program. Thus, with the end of the war, it was demobilized. The National Defense Act of 1920, however, contained a provision that became important to Mines and a great many other colleges and universities. This provision was to redress the lack of commissioned officers evident during the war with a campus-based Reserve Officers Training Corps (ROTC) that through peacetime training would provide a continuous flow of officers into the military. The program provided on-campus training, as well as summer-camp experiences. Some schools

Mines Radio Station
(from the 1921 *Prospector*)

CSM on the air!

To keep up with the trend among mining corporations in the use of radio communication instead of telephones for isolated camps, Mines opened its own radio and telegraph station on 1 January 1921— 9XA1. The United States government licensed the station on 1 February of that year, with C. E. Heffelman '24 operator in charge. Radio communication was first carried on with the South Dakota School of Mines, 450 miles away. The receiving equipment included a Grebe Special Receiver, type RC, with two steps of audio amplification. The sending equipment comprised a one-kilowatt closed core transformer, rotary quenched spark gap, oil-immersed condenser, and oscillation transformer with protective devices of the most modern design. A 100-watt radiotelephone set was also installed. The range of the receiving equipment was 2,000 miles. Signals were distinctly heard from as far away as Honolulu, Hawaii, and Havana, Cuba. The transmitting range was 1,000 miles.

made the first two years of the program compulsory, while others did not. Generally, the small liberal-arts colleges did not participate at all. On the other hand, the act required land-grant universities to make the first two years of ROTC compulsory. For those students entering the third and fourth years, scholarships were provided. For those students commissioned, active duty of some kind was generally required. Even those with only two years of ROTC would, in case of war, be able to move more readily into officer candidate schools. The proof came in 1940, when the War Department activated over 80,000 reserve officers, most of them from ROTC programs.

The Army was the pioneer in early ROTC programs, but eventually the Navy and the Air Force, particularly after World War II, had programs on many campuses. Quick to apply for an ROTC unit, Mines received permission to organize a Senior Division Engineer Reserve Officers Training Corps unit in the fall of 1919 and became one of the first four collegiate institutions to receive this honor.

The goal of the ROTC was, and still is, to educate and train junior officers in a basically civilian academic environment as opposed to a totally military one, the desired end being a cadre of educated and trained officers cognizant of the relationship between the civilian and the military worlds and of the value of leadership in both. Because Mines was not a land-grant school, the government required no years of ROTC, but the Board of Trustees made the first two years a requirement for all physically able students. So, with Major G. C. Dotson as commandant and seventy basic-military cadets enrolled, Mines embarked on a long and rewarding partnership with the U.S. Army Corps of Engineers. Following an inspection of the Mines unit in 1921, Alderson, a strong supporter of ROTC, spoke of the value of the program to colleges and universities:

In my opinion, the Reserve Officers Training Corps is an excellent adjunct to any American college and especially to our technical colleges. The question of discipline, obedience to authority, and loyalty to the nation are elements in an education that are of vital importance today. The work of the Corps brings these elements of a young man's education into prominence. In addition to this, the training a young man gets gives him an added value as an engineer.

On the whole, I feel the Reserve Officers Training Corps is an exceedingly valuable feature of American college life today. My observation of its work at the Colorado School of Mines makes me an ardent advocate of its continuance. (MM February 1974)

Out of necessity, engineering and the military over the centuries have developed a close and practical relationship. Georgious Agricola long ago remarked at the end of Book I of his *De Re Metallica* that miners had a number of characteristics that made them excellent soldiers:

For, trained to vigilance and work by night and day, he has great powers of endurance

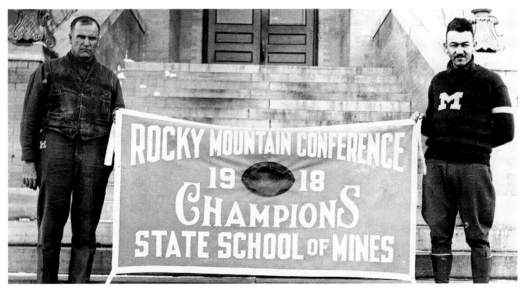

Wave the banner for CSM!

Rocky Mountain Conference champions—1918.

when occasion demands, and easily sustains the fatigues and duties of a soldier, for he is accustomed to keep long vigils at night, to wield iron tools, to dig trenches, to drive tunnels, to make machines, and to carry burdens. Therefore, experts in military affairs prefer the miner, not only to a commoner from the town, but even to the rustic.

A more colorful tribute to the military engineer may be the poem "The Army Engineers," dedicated to the "Fighting Miners" of the Army Engineers. Though somewhat doggerel, it nevertheless speaks the truth, closing with the following stanza:

> For they are the lugs that lug the most
> Of loads that land on a hostile coast,
> And they are the guys, when equipment fails,
> Who scratch out a ditch with their fingernails.
> You find them up in the mountain crags
> And down in the jungle clearing snags

Where the moccasins coil and the snipers lurk
"Mines" Engineers, doing the dirty work!
(MM November 1943)

II

World War I, the "war to end all wars," with its industrialized slaughter of millions of human beings, was a bitter testament to technological advances that leapfrogged human understanding. It was indeed the "impossible war" that the Polish-Russian businessman Ivan Bloch in his multivolume *Future of War* (1898) said could never happen because war was becoming too deadly and too expensive to fight. Along with the all-too-visible loss of human lives came the abstract loss of human values and beliefs, those qualities necessary for a sense of purpose and stability. To paraphrase the poet Wallace Stevens, the old gods were dead, and the new gods had yet to be born. The world did indeed have to redefine itself. Four great monarchies

The baseball team, too, captured a conference title in 1918.

A Plea to the Alumni
(from MM, December 1923)

Another disastrous football year has passed [two wins–five losses]. We bow silently, cuss inwardly and wonder. There has not been a real winning team for years. What is the reason? Can it be answered in a few terse words? There are so many details that, singly, seem unimportant, but are, in reality, the situation in a nutshell, that one dislikes advancing them for a football failure excuse.

Briefly, the material is not here at Mines. The Alumni must shoulder the blame for this failure. There are many skilled athletes playing in the vicinity of a Miner's home, who are not learning of the Colorado School of Mines. You've got to talk, MINES, MINES, MINES, before the men come to Golden. There are no special athletic inducements. State scholarships can be awarded, but that is the limit. Other schools pay; Mines does not. Our row then, with a difficult curriculum, is difficult to hoe.

Players don't have enough time to practice. Classes are not dismissed until 4:15 and the men are not working until 4:45. Coaches are plentiful. But—tho we have an athletic director to supervise the business end of the athletics, we have NO ONE MAN TO TAKE GENERAL CHARGE OF COACHING ALL THE YEAR ROUND. Such a man is needed. We cannot continue the practice of having a coach on the job three hours a day for three months—only to stay away the other nine months and probably fail to return the next year.

The caliber of men is changing. Mines men are primarily students, are younger than Miners of fifteen years ago, and are not as husky.

Too many difficult subjects, no inducements for athletics, not enough time for practice, the lack of a general head coach, the changing student, not enough athletic material, and the failure of Alumni to talk MINES are reasons why Mines is a laggard in football.

The reasons are simple to discern, but the remedy? Alumni can do their part by sending the athletes to Golden. When that cooperation is obtained, the rest will be worked out. But the Alumni have the most difficult task of all. Are you ready, Alumni, to help Mines football? If so, line up the athletes for next year. We need them to regain the athletic prestige that once was Mines'.

---

were destroyed; two major democratic empires were weakened; new nations were created; and new problems, not to mention another war, loomed on the horizon.

Although the United States did not suffer from the war to the extent of other participants, it did undergo rapid change as a nation. Mass production and concern about managing the "machine," along with the desire for social justice, brought an increase in both bureaucratic agencies and in labor union activities. The natural growth of population was bolstered by legions of immigrants (almost 15 million between 1901 and 1920), most of whom settled in larger cities. With prosperity and

Members of the prospector's course in the chemistry laboratory.

optimism the mode, America welcomed the "Roaring Twenties," a more-or-less screwball compound of religious revivalism, prohibition, bootlegging, gangsterism, women's rights, radio, silent movies, raccoon coats, the "flapper," increased interest in sports, and the closed automobile (a kind of motel on wheels). These were the ingredients of American culture during Victor Alderson's remaining six years at Mines and for the remainder of the decade.

Whatever he and the Board of Trustees hoped it would be, Alderson's second term as president was certainly not without dissension. There were still elements of the faculty, the student body, and the alumni who found it difficult to work with him and who disagreed with his apparent dictatorial manner, as well as with his goals for the school. There were enough specific incidents of conflict to keep unrest simmering, if not always boiling. Still, during Alderson's second term, Mines did make positive strides as it moved further into the twentieth century. With the exception of the blip in enrollment that occurred during 1917–1918, the student body was growing, as was the faculty. Mines' reputation, too, continued to place the school among the nation's leaders of mining-engineering education.

No matter how many times flags of dissension have waved over the Mines campus, it is important to remember that the progress of the school has generally not been in spite of its leadership but because of it. Progression and regression, moreover, are not always easy to define in the short term, not to mention that, like beauty, each is in the eye of the beholder. Furthermore, it is important to keep in mind that a college or university president, no matter how strong or weak, is not really the lifeblood of an institution or the sole director of its progress or lack thereof. When the chips are down in the game of academic poker, it is the faculty members teaching students and doing their research that carry a school through both calm and turbulence. What a president can do, as the occasion demands, is serve as a stimulus, as a guide, as an antagonist, as a fundraiser, and, in the words that Robert Frost used to define a poem, as "a momentary stay against confusion." Generally, the presidents of institutions, like the leaders of nations, rightly or wrongly, get the credit or the blame for anything that happens on their watches. This is perhaps because they usually make the most money

The Whole Damn Faculty.

and have the most power or because they are simply "sitting ducks." It is no exaggeration to say that presidents must try to be all things to all people, be they faculty, alumni, students, trustees, or financial benefactors. In a lighter note on the topic of being a president of a university, Clark Kerr in *Time* magazine of 17 November 1958 said, "I find that the three major administrative problems on a campus are sex for the students, athletics for the alumni, and parking for the faculty."

In some respects, being president of a small college is perhaps more difficult than being president of a large university where a protective bureaucracy can insulate one from the day-to-day squabbles that exhaust both time and energy. In the academic game at any level, however, it is wise for a president not to take himself or herself too seriously and, above all, not to fall into the thought-trap that one anonymous president did when he said, "If I didn't have to put up with the faculty and all these students, I could really run this school."

III

Mines entered the 1920s as a promising institution rapidly leaving behind the nascent school of earlier

Brooks Field at rest.

years. By 1922, it offered eight degrees: Mining Engineer, Metallurgical Engineer, Chemical Engineer, Geological Engineer, Petroleum Engineer, Mechanical Engineer, Electrical Engineer, and Civil Engineer. Dr. F. M. Van Tuyl, who joined the faculty in 1917, became head of the Department of Geology in 1919, a position he held until his retirement in 1953, establishing a record thirty-six years of service. Under his leadership, the department developed a strong curriculum and ultimately offered the nation's first degree in Geological Engineering. As Van Tuyl conceded in an unpub-

Blue Key

A group of Mines students of the early 1920s formed a secret society called the Vigilantes or the "Vigs." A watchdog group, their purposes were to promote Mines spirit and to advocate the ideals of college life that would be of credit to both students and school. Their activities, however, were often spontaneous and of a more-or-less "good-fun" nature: painted slogans, dynamite blasts the morning of a football game, midnight haircuts, one-way rides into the mountains for those not living up to the Mines spirit, and whatever else might strike their fancy.

From some members of the Vigilantes came the idea of forming a chapter of the Blue Key National Honor Fraternity, founded at the University of Florida. On 29 January 1926, "Vig" members and representatives from fraternities and the Barbs met at the Berrimoor Hotel to form the Vigilante Chapter of Blue Key. With the motto of "Serving, I live," Blue Key grew to prominence on the Mines campus as a group dedicated to the well-being of student and school and to the preservation of, and participation in, campus activities, particularly in the freshman year.

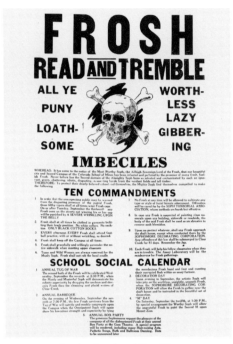

Frosh—Read and Tremble.

lished 1965 memoir titled *Confessions of a Restless Geologist* (Archive), there was considerable prejudice against such a degree among geologists because many believed that "geology was a qualitative subject rather than a quantitative subject," though, according to John Haun, a more recent Mines geologist, "The extent of 'prejudice,' if it ever existed, was not based on qualitative versus quantitative aspects of geology, but on presumed lack of academic rigor in the dirty business of engineering practice." Van Tuyl observed that as time passed, "the development of many quantitative applications of geology, particularly in engineering geology and in mining and petroleum exploration and exploitation have justified the awarding of such a degree. Eighteen colleges and universities now offer curricula in this field."

The catalog of Alderson's last year as president (1924–25) offers further evidence that Mines was advancing. The enrollment had grown to 535, with 20 postgraduates, 118 seniors, 97 juniors, 137 sophomores, and 163 freshmen. Why the junior class was smaller than the senior class is unclear, other than perhaps an unusual attrition rate between the second and third years. But, most importantly, the enrollment had increased by 76 students from the previous year. The faculty numbered forty, having twenty-one visiting special lecturers and nine staff members. Student organizations included, among others, the Student Council, the Band (forty pieces), the Glee Club (twenty-five members), the *Oredigger* (the school paper started in December 1921), the Junior Affiliated Society of the American Institute of Mining and Metallurgical Engineers (the largest junior affiliate in the nation at over two hundred members), the American Association of Engineers (student chapter), Radio Station 9XAI (founded on 1 January 1921), Tau Beta Pi (honorary fraternity), Theta Tau (professional fraternity), Sigma Gamma Epsilon (professional fraternity), the *Prospector* (yearbook), and five social fraternities—Beta Theta Pi, Kappa Sigma, Sigma Nu, Sigma Alpha Epsilon, and Mu Epsilon Tau.

The campus boasted nine instructional and research buildings: Hall of Engineering, Guggenheim Hall, Hall of Chemistry, Hall of Advanced Chemistry, Assay Building, Gymnasium, Stratton Hall, the Armory (leased from the state), and the Experimental Plant. The Armory joined the campus

1920 baseball team.

buildings in 1922, providing dormitory space, an infirmary, and a banquet room with kitchen. The president's house, the Power House, and two shop buildings made up the remainder. Brooks Athletic Field, named in honor of Ralph D. Brooks, a member of the Board of Trustees, was built in 1922. It was comprised of a baseball diamond, a quarter-mile track, a one-hundred-yard straightaway track, a football field, and a grandstand that seated 2,500. Important to the alumni and to the school as a whole was the first Homecoming Day celebration on 15 November 1924. Complete with a parade, a football game, and a banquet, it was indeed a gala event from morning through the night.

Mines tangles with archrival University of Denver.

These features, along with its growing reputation in engineering circles, designated Mines an institution

"Rut" Volk

The gymnasium that was completed on the Mines campus in 1960 had no name other than simply the Gymnasium until the evening of May 9, 1974, at which time it was dedicated to, and renamed for, a man considered by many as the greatest Mines athlete of his generation: Russell H. Volk '26. The memorial portrait displayed in the Russell H. Volk Gymnasium shows a man of trim build with a partial smile on his handsome face. The subject of a portrait, of course, cannot leap from its frame to illustrate the athletic prowess that earned him such renown. That remains for the record books and the memories of the dwindling few who may have seen him on the football field, the basketball court, the baseball field, the wrestling mat, or in the boxing ring.

During his career at Mines, "Rut" earned a record fifteen varsity letters in intercollegiate sports: three in football, one in basketball, three in baseball, four in boxing, and four in wrestling. He won all-conference honors five times. He was the only athlete to win the conference 175-pound wrestling championship and the heavyweight boxing championship—on the same evening.

In addition to his athletic achievements, "Rut" served as president of his class and was national chairman and Mines delegate to a student leadership conference in Yugoslavia.

Completing a master's degree at Mines in 1931, "Rut" built a career in the petroleum industry that led to his founding of the Plains Exploration Company, a production firm that pioneered in petroleum exploration and development. A noted civic leader, "Rut" was appointed by Governor John A. Love to the Mines Board of Trustees, a position in which he served from 1967 to 1973. He also served as president of the Mines Alumni Association and as director of the Colorado School of Mines Foundation. "Only one 'Rut' Volk was made," said Colonel Wendell W. Fertig at the dedication dinner, "and the world and Mines is a poorer place now that he is gone."

The ROTC on parade.

that was not merely marking time. Even with these advances, there was still no peace in the valley. Clouds of contention hovered over the campus as the feud between Alderson and the Board of Trustees on the one side and a number of alumni and various faculty members on the other waxed with virtually no waning. The various aspects of this conflict were complex and reflected many personal animosities, as well as philosophical disagreements as to what Mines should be and what duties the president and the trustees should perform *vis-à-vis* the faculty, the students, and the alumni.

As might be expected when a school grows and begins to expand and redefine its mission, those who know the school as it once was, particularly alumni, are prone to question the wisdom of change. The president and the trustees are often the prime movers of such change. The faculty's role depends on the dynamics of the relationship that its members share with the administration—though, of course, the faculty ultimately has to implement and carry out any change. That Mines perforce must change to meet the times, then, was one factor. The reappointment of Alderson to the presidency over the objections of a number of alumni and current and former faculty members, as we have seen, was another. Yet a third factor was the question of who has the ultimate responsibility for operating a college—the president, the trustees, the faculty, or the students. These factors combined to ensure that more than dynamite would soon explode on the Mines campus.

In 1917, three former faculty members, two of whom had been allegedly dismissed with neither notice nor cause and one of whom had resigned, requested that the American Association of University Professors investigate Mines' treatment of the two dismissed professors, as well as the seemingly constant turmoil on the campus. The following were the charges:

1. That the Board of Trustees intervened in matters of student discipline.

2. That President Parmalee's dismissal was chiefly due to his stand in resisting the attempt of trustees to intervene in the question of discipline which arose at the time of the students' strike.

3. That a number of teachers of professorial rank were dismissed, mainly during the administration of President Alderson,

without charges, hearing, or adequate warning.

4. That President Alderson discouraged high standards of scholarship and brought pressure to bear upon instructors to raise the marks of students.

5. That President Alderson used his power of reappointment to secure support among the faculty for his personal policies.

The exhaustive investigative report appeared in *Bulletin of the American Association of University Professors* of May 1920 and covers the testimony of both sides and the Investigating Committee's findings. It also provides a fascinating view of the conflicts that had erupted among the constituents and an insight into some of the general problems facing a growing collegiate institution as it attempts to define and differentiate the responsibilities of faculty, trustees, students, and even townspeople regarding the school's daily operation. While the twenty-one-page report is well worth reading in its entirety, space here permits only a brief recounting of its conclusions. A condensed version of the report may be found in MM December 1920.

According to the *Bulletin*, war conditions delayed the Investigating Committee's collating evidence pertaining to the charges of the complainants until November 1918. Further delay occurred because three of the five members of the Board of Trustees at

The Beta Theta Pi house.

It looked like a mini castle then, and it still does today.

Mines were concurrently completing their terms, and the committee decided to wait until new appointments brought the board to its full complement of members. Even then, some months passed before the board indicated a willingness to reply to the charges. The new board initially argued that it was not responsible for the actions of previous administrations/boards. By late 1919, however, the board, after reviewing the charges and the compiled evidence, agreed to respond.

The board's refutation of the charges consisted of the following:

1. The tenure of service in the faculty is fixed in the resolution of employment. When a modification of this modicum of employment was suggested in 1915 by President Phillips, the faculty argued for the status quo.

2. What a member, or members, of the Board of Trustees may do as an individual or individuals is not binding upon either the board or the faculty, and constitutes no charge against the Board of Trustees as an organized body. Moreover, even if the board did meet with groups of students, citizens of Golden, and state officials regarding the school, such would be no violation of the jurisdiction of the faculty.

The Integral Club

The Integral Club was a favorite gathering place for the Mines community.

President Alderson and the alumni association organized the Integral Club, better known as simply the I-Club in more recent times, in the fall term of 1909. With rooms in the new gymnasium, which came a year earlier, it was hailed as another sign that Mines was becoming an up-to-date school. The club, run by a committee of students, was open to students, faculty, trustees, and alumni, as well as to subscribers to the Gymnasium Fund. The main clubroom was well appointed with pool and billiard tables, a piano, smoking tables, chairs, a sofa, and a confection/cigar stand. The walls were decorated with pennants and pictures, and the mantle above the large fireplace held various trophies celebrating Mines athletic accomplishments. Footballs from significant victories hung from the ceiling. These victories were aided by rallies held in the club before games. Meetings or rallies held elsewhere could not match the intensity of the Mines spirit evident at the club rallies. The room became dear to the hearts of all Miners. A second room was set up for games such as chess, checkers, cribbage, and other card games. Hours were from 9:00 a.m. to 11:00 p.m. on Sundays through Thursdays and from 9:00 a.m. to 12:00 midnight on Fridays and Saturdays.

The I-Club still exists today in the Student Union though its mellow and intimate atmosphere has diminished over time. Its primary function now is to serve cafeteria-style breakfasts and lunches. It is no longer open in the late afternoons or evenings except for special events.

3. Dismissing President Parmalee was entirely within the powers of the board.

4. The dismissal of the professors in question was based in one case on neglecting classes in order "to perfect an automobile and other mechanical devices" and in two cases on the lowering of the standards of the school by giving high school level courses in algebra and solid geometry.

The Investigating Committee concluded that Alderson was responsible for the dismissal of capable teachers without hearing or adequate warning and that he lowered the standards of scholarship by compelling faculty members to change student grades, especially of those with influential relatives. Regarding the Board of Trustees, the report stated that individual members interfered with matters of student discipline and encouraged students to flout authority, particularly in the case of the dismissal of President Parmalee. The report also verbally chastises the Board of Trustees for failing to understand and carry out the various responsibilities delegated to them by the people of Colorado. Specifically, the team adversely criticized the following resolution of the Board of Trustees that in their view had a negative foreshadowing:

> Resolved, that loyalty to the institution, to its ideals and to those connected with it, will be considered of the utmost importance, and that even though ability and attention to duties are marked, any member of the faculty or other employee who attempts to undermine the school or to advance personal interests at the expense of his colleagues or the school will not be retained.

Summarizing, the Investigating Committee stated that unless the school remedied the situation that caused the investigation, it would not be able to attract and retain teachers of superior qualifications, which would be a disastrous result for the institution.

Nor would the AAUP investigation be ignored by the scientific and technical journals. The

The band was a vital part of the Mines scene though not in hard hats and jeans.

Homecoming arrives on the Mines campus.

Engineering and Mining Journal bluntly advised the people of Colorado that "no mining school in the country has for years borne so dubious a reputation as being the playground for politicians, as has that of Golden. This must mean a good deal, for we have no doubt that much of this evil exists in other state mining schools; but at least nowhere else have we had it so repeatedly thrust on our attention that it reeks to heaven" (31 July 1920). The article further noted that like California, Colorado "is the mother of miners. It has one of the most advantageous locations on earth for a mining school, and its school should rank among the highest. It does not." The *Mining and Scientific Press* came down just as hard on Mines with a list of its internal problems: "In 1897, 1902, 1903, 1905, 1915, 1917, and in 1920, the School suffered from unfortunate publicity arising from domestic troubles. Sometimes it has been a president who has shown no decent consideration for the members of his faculty; sometimes it has been gross interference by the trustees with the authority of the faculty; and sometimes, in consequence, it was the students who proved mutinous" (11 September 1920). Both journals urged the school to divorce itself from local politics, particularly in the way that trustees were chosen. Although admonitory in tone, both articles expressed a sincere hope that the school would be able to salvage its reputation as a leading institution of mining engineering. Two factors are important here: (1) Mines started out with strong support from the people of Golden, who in turn felt a vested interest in the school; and (2) the territory and the state of Colorado were both essentially sparsely populated at the time and not far removed from frontier days. These factors probably militated toward greater political pressures than would have occurred if the school had been founded in a more populous area of the East.

The AAUP from its inception has championed academic freedom and the rights of faculty members, particularly in terms of tenure and dismissal. Unlike accrediting agencies such as the North Central Association or the Accrediting Board of Engineering and Technology, the AAUP has only symbolic power. A number of schools, for example, appeared on its censure list for years with little negative impact, other than the occasional necessity to explain the situation to a prospective faculty member who might be curious. Most students and their parents, on the other hand, have little if any knowledge of the AAUP and its mission or the concepts of academic freedom and tenure. Still, most collegiate institutions want to avoid being symbolically ostracized by any major academic association. For Mines, the 1920 AAUP investigation would not be its last. Indeed, seventy-four

Not exactly a heavyweight match, but a match nonetheless.

Whose Song Is It, Anyway?

It has been said that if you ever need a friend virtually anywhere in the world, simply sing a verse or two from "The Mining Engineer," and you will soon have one—or two. But will the friend(s) be from Mines or from Georgia Tech? Based on "The Son of a Gambolier," a spirited musical biography of an Argentine (some say English) soldier of fortune, the song in question has been a "cause célèbre" between Mines and that other school in Georgia regarding which had the song first. Over the years, each has marshaled evidence and presented arguments to prove rightful ownership.

Mary E. Hoyt, many years the librarian of Mines, in her short history of Mines (MM June 1949), traced the song at Mines back to the late 1870s when the school's Glee Club sang it at its popular programs. That would pre-date Georgia Tech's claim since it was not chartered until 1885 and did not begin instruction until 1888. The question here is whether Mines actually had such an early glee club. Other sources point to 1911 as the organization of the first glee club.

On the other hand, the *Oredigger* of January 3, 1928, maintained that Georgia Tech did indeed bring the song to America and that Mines picked it up from those "Rambling Wrecks." Like so many mysteries, this one is perhaps best left to speculation; the song is big enough for both institutions. As for "The Mining Engineer," it has spoken eloquently for itself and for Miners everywhere —as it certainly did for the graduating class of 1895 as they sang it marching into the Methodist Church for their commencement exercises.

> I wish I had a barrel of rum and sugar three hundred pounds.
> The college bell to mix it in and clapper to stir it 'round.
> Like every honest fellow, I take my whiskey clear.
> I'm a rambling wreck from Golden Tech, a helluva engineer.
>
> A helluva, helluva, helluva, helluva, helluva engineer.
> A helluva, helluva, helluva, helluva, helluva engineer.
> Like every honest fellow, I like my whiskey clear.
> I'm a rambling wreck from Golden Tech, a helluva engineer.

years would pass before Mines would be in full compliance with AAUP guidelines.

C. Erb Wuensch, a journalistic firebrand and ardent Mines alumnus, as well as the editor of MM at this time, expressed his own views in an editorial urging the school's alumni to unite to bring about needed changes. He recommended that all state schools be put under the jurisdiction of one board of regents to prevent vested-interest interference at the local level (MM December 1920). In the Discussion section, a number of alumni letters for one reason or another agreed with Wuensch's views. One letter with a tint of unintended foreshadowing suggested a change-of-pace plan: "My plan would be to put Mines under the control of the Greeley Normal School. Just think what a refining influence the Normal girls would have on our School! In time that rough yell: 'Give 'em Hell, Mines' would be abolished and we might even hope to have a lady for president of the School" (MM January 1921).

Physical Development for Miners

Guess who! The Mines swim team.

Ralph Glaze came to Mines as football coach in 1919 with a reputation for making coaching a profession rather than a side issue. A graduate of Dartmouth, he was named to Walter Camp's All-American team at end. He also pitched for the Boston Red Sox. He brought with him a plan to popularize physical development among Mines students "so that a man would know how he stands physically." Below are the goals to be attained for the Honors Division (the highest of three):

1. HEALTH. Sound heart and lungs. Satisfactory nerve and blood conditions. Normal conformation and carriage. A clean bill of health.

2. AGILITY. Ability to hand-vault an obstacle the height of the eyes. Ability to run and jump obstacles the height of the nipples. Ability to broad jump eighteen feet. Ability to run one hundred yards in eleven seconds.

3. DEFENSE. Ability to execute eight wrestling maneuvers. Ability to demonstrate the fundamentals of boxing or fencing.

4. SWIMMING. Ability to swim a quarter-mile. Ability to swim twenty yards supporting a helpless companion. Ability to dive from a height of ten feet and bring a helpless victim to the surface. Knowledge of five strokes and ability to swim twenty-five yards in fifteen seconds. Knowledge of first aid.

5. SPORTSMANSHIP. Varsity squad in at least one sport. Reputation for sportsmanship and courtesy.

Wuensch's views did not go unnoticed. In a letter postmarked 17 January 1921, the Board of Trustees advised him that it would no longer provide financial support or headquarters for the magazine "until such time as the Magazine shall loyally support the school and fulfill its obligations to the institution and to its alumni." The administration changed the locks on the office in question, and Wuensch was forced to move to Denver. In that same month, he submitted his resignation as editor of the magazine, admitting that he had departed from his previously avowed editorial policy of neutrality with the goal of stimulating alumni action toward improving Mines (MM January 1921). William S. Simpson, secretary of the alumni association commented, "It has never been and is not now the policy of the Executive Committee, which is responsible for the Colorado School of Mines Magazine, to employ the columns of this publication in the manner appropriated by Acting Editor Link's immediate predecessor [Wuensch]" (MM April 1921). He emphasized that the alumni association still believed that all state colleges and universities should operate under one board of regents.

Given all of these activities, the General Assembly decided to conduct its own investigation into the Mines situation. It appointed a committee to look into the five charges of the AAUP team. In a nutshell, this committee refuted each of those charges and stated that the members of the student body "give to the School of Mines and its administration hearty endorsement and approval." The committee recommended that there should not be a local member on the Board of Trustees and that there should be a fuller recognition of the mining industry on the board. The committee concluded "that the management and administration of the School of Mines is efficient, the Trustees, officers and faculty competent, well qualified and trustworthy, and that the Institution, members, officers, faculty and Trustees are entitled to the support, respect, and encouragement of the citizens of this State, the alumni of the institution and the general public." This statement essentially covered all bases. The two conflicting views of the AAUP investigative team and of the General Assembly's investigative committee indicated that the truth probably lies somewhere in the middle. (See MM March 1921 for a more complete discussion of this

A number of fraternities had a dog. Here is the SAE chapter and their "pooch."

Yes, Mines students could dance in the 1920s.

question.) At the same time, it should not be surprising that the General Assembly's committee would defend its own school against outsiders.

Although Alderson was given a three-year contract as president in 1922, it was only a matter of time before Mines would be looking for another president. The school certainly could not let go uncorrected the kinds of activities that invited a negative report from the AAUP and negative editorials in highly respected and widely read journals. Some action had to be taken. So, once again hoping to clear the air of dissension, the Board of Trustees declared the presidency vacant as of 1 September 1925. Alderson was gone once more. In all, he served some eighteen years as the administrative leader of Mines, second only to Melville F. Coolbaugh, his replacement, in total length of service. In spite of Alderson's being a lightning rod of dissent, when the dust of controversy was swept away and the results tallied, the school did fairly well under his leadership. Even though it had its detractors, the revised curriculum appeared to fit the school's mission and to strengthen its standards. Moreover, buildings were built, and the enrollment of 475 (first semester of 1924) dwarfed the enrollments of the other twenty-two institutions that offered mining engineering. It was an enrollment that drew from high schools and academies and from other colleges and universities, a fair number of them foreign. It might be said then that Alderson served Mines well—well enough to be awarded an honorary Doctor of Engineering degree in 1938 and to have a building named for him in 1954, Alderson Hall (now housing Petroleum Engineering and Chemical and Petroleum Refining Engineering). Strange indeed are the ways of academe.

Max W. Ball, a Mines alumnus and a new member of the Board of Trustees, in an address to the Colorado Metal Miners on 22 January 1924, presented an overall optimistic view of the status of the school. Four of the five members of the board had been appointed within the past eight or nine months, he said, and two members now represented metal mining. In terms of its curriculum, Mines had "once more become exclusively a mining school" and the Experimental Plant, a target of much criticism for a number of years, was now functioning under new leadership and was "ready to tackle practically any metallurgical or mining problem." Hence, with a new Board of Trustees and with a new president about to appear, Mines was "sitting on Go" and ready for the next toss of the academic dice.

IV

When he came west from an Iowa farm to attend Colorado College, Melville F. Coolbaugh probably was unaware of the intense athletic rivalry between his alma mater and Mines. Nor in all likelihood did he picture himself as a future president of that mining school up the road a piece. After obtaining his Bachelor of Science degree from Colorado College, he went east to Columbia University, where he received a Master of Science degree. Thus equipped, he came west again, this time to the South Dakota School of Mines in Rapid City where he taught chemistry. Just prior to America's entry into World War I, he came to Mines as a chemistry professor, eventually becoming the department head. After serving in the war, he went to work for the Metals Exploration Company in Denver. There he developed a process for the recovery of ores—the Coolbaugh Process. He then formed a subsidiary company of Metals Exploration, with himself as president. It was then that the Board of Trustees offered him the presidency of Mines. Since Coolbaugh, when he was previously a faculty member at Mines, had been a part of the opposition to Alderson, it might have surprised some people that the board would choose him as president. But they no doubt saw him as a leader who could bring the faculty together and, at the same time, restore the alumni's confidence in their alma mater. The board was right on both counts. If Regis Chauvenet took Mines through the final stage of its formative period and beyond, Coolbaugh surely led the school through twenty years of stability and advancement, all without the specter of unrest that had so often haunted the school. His administration was to be a well-needed defining period in the school's history.

Dr. Robert Weimer, internationally renowned Mines professor emeritus of geology, in his

---

E-Day

Many colleges and universities have a day (or days) in the spring during which the student body celebrates the end of winter and the approaching end of the spring term with various activities peculiar to individual campuses. That day for Mines started in 1927 as a joint effort between the Colorado Engineering Council and the school. Activities included technological exhibits, campus tours, a dance, and a basketball game. Since then, it has developed from a relatively small event to one that rivals homecoming on the Mines calendar. In 1939, Mines and the council gave the students the responsibility of planning and carrying off the day. With America's entry into World War II, E-Day celebrations were canceled for the duration. Following the war, students sought to expand interest in the day by adding new activities and components: an E-Day queen, a drilling contest, a mucking contest, a regatta, and even class reunions. Because of increased community interest, a second day of celebration was added, with guest lecturers, band and glee club performances, and demonstrations of heavy equipment. As the budget grew, so did E-Days. Fireworks, popular musicians, symphony concerts, and numerous other activities were added. Still the most noted, however, is the ore cart pull to the capitol. Traditions have come and gone at Mines, but E-Day as a reflection of modern-day Mines spirit is here to stay.

Thirty-nine gas-fired muffle furnaces in the assay building provide means for practice in assaying of gold, silver, and lead ores.

presentation at the inauguration of incoming Mines president George Ansell, on 8 February 1985, made an accurate and telling division of the history of Mines: the first 51 years and the last 60. He stated that the "first 51 years seemed to have cycles of unrest and tranquility. Times of discontent were caused by conflicts among the faculty, administration, Board of Trustees, alumni, athletic programs, and unrest among students. That covers about everyone." The last 60 years he described as primarily a period of tranquility and progress. Dividing by two in order to characterize the history of Mines is a big sweep, but Weimer was on target in terms of the big picture.

The challenge for Coolbaugh, as he well knew, was to lead an institution that had passed through a number of controversial issues to a new level of development. Thanks again to a faculty that for the most part carried out its academic responsibilities with acumen and dedication through a tumultuous period, Mines was ready for Coolbaugh's leadership for the next twenty-one years. In his inaugural address, Coolbaugh packed an entire philosophy of education into a few well-chosen words. He deftly divided the forces of modern education into the dichotomous extremes of conservatism (or standardization) and radicalism (or change). "We are," he said, "ever swinging back and forth between the extremes, trying to find the most potent proportion between them. Youth in general is radical, old age conservative." To the reactionary, the past is always better than the present, let alone the future. Indeed if the forces of reaction had prevailed through the ages, "we would still be ploughing with a crooked stick." On the other hand, he noted, the radical view is just as simplistic as its opposite: "Good is to be found only in the future. Continued change is its program, since any newly established condition becomes old immediately and is therefore naturally abhorred by the radical. Continual change without reason can produce nothing but chaos." Coolbaugh emphasized that underlying principles and fundamentals must be guarded at the same time that individuality and creativity are encouraged. Nevertheless, his summation presents a challenge for educational institutions of any time: "To train men and women who recognize the value of the past and build upon it the greater structure of the future should be the ambition of every educational institution. Learning not for learning's sake, but for living, and understanding, and growing should be the ideal of every system of education" (MM October 1925).

Coolbaugh's first order of business in terms of curriculum was to reduce the number of elective courses offered and to discontinue the degrees in Chemical Engineering, Civil Engineering, Electrical Engineering, and Mechanical Engineering. All of these deletions were effected in 1926. The remaining degrees included Mining Engineering, Metallurgical Engineering, and Geological Engineering. This particular adjustment was welcome news to those alumni who felt that the school's curriculum had been straying too far from its forte of mining engineering. Also in 1926, Coolbaugh persuaded the Board of Trustees to initiate courses in Geophysics taught by Dr. C. A. Heiland. In 1929, a fourth degree, Petroleum Engineering, was added to Mines' offerings. Also in 1929, a graduate school was created that offered a Master of Science degree and a Doctor of Science degree.

An editorial from the *Engineering and Mining Journal* applauded Mines after its editor T. A. Rickard had attended the June 1927 Mines commencement. After praising the dignity of the ceremony itself and the addresses by President Coolbaugh and William H. Smiley, president of the Board of Trustees, Rickard contrasted the Mines of 1927 with the Mines of 1917:

> Ten years ago, the Colorado School of Mines was in a bad way; indeed, it has had more than one dip in its upward curve. Today, besides its efficient president, the school has a board of trustees that works loyally with the faculty and commands the respect of the people of Colorado. Dr. William H. Smiley, formerly head of the public schools of Denver, is the chairman [president] of the board of trustees; he is an experienced educationist and a much-respected citizen. With him are Mr. Benjamin F. Hill, a graduate of Columbia and a geological practitioner of assuring standing; Mr. Robert H. Sayre, a mining engineer and the son of a mining engineer, a graduate of Harvard; Mr. William A. Way, a lawyer ready to steer his colleagues amid the shoals of legal procedure; and Mr. Max W. Ball, the secretary, a petroleum engineer, an alumnus of the school, a man whose engaging personality and keen mentality have made him well known throughout the West. Evidently this is a good working board of trustees, with enough technical skill to make them real helpers to the president and the faculty. (9 July 1927)

This testimony to the members of the Board of Trustees was welcomed by all at Mines: administration, faculty, students, alumni, and board. It was particularly uplifting, given the negative editorials of previous years regarding the trustees and faculty of Mines that had appeared in the *Engineering and Mining Journal* and other influential engineering journals. That was the past, and this was now. Plaudits were not only welcome, they were deserved. With the clouds of dissension disappearing behind the mountains and with a capable and dedicated president working in harmony with a strong faculty and a restructured Board of Trustees, Mines was ready to advance on all fronts.

New clouds, however, appeared with the 1929 crash of the stock market and the ensuing Depression that gripped the nation. The 1929–30 catalog lists sixty-four faculty members and ten staff members. The student enrollment totaled 423—16 graduate students, 64 seniors, 99 juniors, 115 sophomores, and 129 freshmen. Of the total, 242 were Colorado residents, with the rest spread widely across the United States and eleven foreign countries. This number was considerably lower than the 535 total in 1924–25 and even slightly lower than the 1922 total of 459. It is difficult to ascertain the exact effect of the crash on enrollment, but it must have been considerable. Although the road ahead was not easy, it was at least one that Mines could face with a sense of unity and purpose.

CHAPTER 5

Reaching Out

Only that day dawns to which we are awake. There is more day to dawn. The sun is but a morning star.
—Henry David Thoreau

When the stock market crashed in 1929, most Americans probably had no idea what or where Wall Street was. By the 1930s, however, few were untouched by the economic chaos sweeping over the United States and virtually the entire Western world. More than the American Dream was being blotted out by business collapses, bank failures, unemployment, drought, and often starvation. The cosmic optimism of progress and perfectibility, so long the basic underpinning of American development, gave way to fear and uncertainty regarding the future of the nation. The free-market concept of laissez-faire economics that had brought America so far and that had given the roar to the 1920s, now elicited only a whimper of despair. Evangelistic preachers and iconoclastic writers pointed to the nation's need for spiritual and material change. Moreover, the increased emphasis on unionism and the specter of communism added to the economic and social malaise facing the nation. In the midst of this turbulence was President Herbert Hoover, one of the most dedicated and fluent supporters of the concept of American individualism and the economic and social systems that it nourished. Not too surprisingly, he was the center of blame for the crash of the stock market and the ensuing Depression, though in retrospect there was probably little he could have done to head off the economic forces steering the world to a vortex of financial chaos. Even though Franklin Delano Roosevelt soundly defeated him in the 1932 election, Hoover never doubted Americans would rise to the challenge and bring the nation back to prosperity and security. Ironically, Roosevelt's New Deal programs and the coming of World War II proved Hoover correct. The country prevailed though the Depression left on the American psyche an indelible imprint that has lasted to this day, particularly on the waning generation of "Depression babies."

Though certainly not immune from the impact of the Depression, most colleges and universities survived the 1930s in better shape than many businesses. The product of a college is education,

Opposite page: A geology class.

Melville F. Coolbaugh begins a long and fruitful tenure as Mines president.

which, though more abstract than the product of a steel mill, a mine, or a wheat field, was still highly valued in most quarters. Despite its initial decline in enrollment and some cash-flow problems, Mines was able to advance, if somewhat carefully, through the economic pitfalls of the Depression toward the dawn of recovery and prosperity.

I

Soon after he assumed the presidency of Mines, Coolbaugh recognized the necessity for the school to increase its resources, financial and otherwise. While support from the state was welcomed and necessary, it was not enough to maintain and strengthen Mines' position in the forefront of mineral-engineering education and research. As it is in many endeavors, the status quo in higher education has ever been a barrier to progress and in some cases even to existence. With these realities in mind, Coolbaugh, in an address at the annual banquet of the Mines Alumni Association in May 1927, noted that the status quo would not permit Mines to meet the increasing demands of the mineral industries. His proposal was a foundation to raise endowment funds for various educational purposes.

The Mines Alumni Association has always viewed itself as more than merely a social organization. The school's concerns were the association's concerns, often to the point of disagreement. The goal for both school and alumni has always been to build a greater Colorado School of Mines. And here, indeed, was a financial engineering problem. Endowments were nothing new to Ivy League schools or to a number of the larger public and private institutions. Harvard, as it does today, set the standard in the 1930s with its endowment of over $100 million. In 1929 alone, Harvard gathered in some $13 million, not to mention annual alumni subscriptions. While they realized such numbers would not be attainable for Mines, Coolbaugh and the alumni firmly believed that the principle of an endowment was valid. Their strategy was to create a foundation that would exist for the sole purpose of advancing the school.

Led by Chairman Charles N. Bell '06, along with Alumni Association President Charles M. Rath '05, a committee began in the fall of 1927 to bring the idea of a foundation to reality. Working untiringly, Bell and Rath, along with their committee, managed incorporation of the Colorado School of Mines Foundation in less than a year, a solid example of the Mines spirit at work. The difference between

Coolbaugh and some of the Mines faculty.
(PIONEER MUSEUM)

Silver Diplomas

When Charles A. Hull engraved a silver diploma for his son, who graduated from Mines in 1933, he probably had little idea of the tradition that he was starting. Intrigued, President Coolbaugh asked Hull if it would be possible to make such diplomas for an entire graduating class. Hull said it would be. With approval from the faculty, trustees, and the state, silver diplomas were granted to each Mines graduate at the May 18, 1934, commencement. Measuring 5" by 6", the diplomas were etched on sterling silver in an elaborate process requiring two men spending five hundred man-hours (roughly six weeks) to complete nineteen delicate operations for each diploma (see the *Oredigger* of March 29, 1963).

Following the replacement of the Professional Engineering degree with the Bachelor of Science degree (1968), the school decided in 1970 to offer silver diplomas to graduate students only and parchment diplomas to undergraduates. In the spring of 1980, the administration indicated that the upcoming graduation would be the last for any silver diplomas, given the inflationary costs of silver. To save the silver diploma, Ralph L. Hennebach '41, president of Asarco, Incorporated, the largest miner and refiner of silver in the United States, said Asarco would provide the silver if Mines would continue to offer the silver diploma to professional and graduate degree students. Mines agreed.

In December 2001, however, the tradition was totally revived. Graduate students continued receiving silver diplomas, and undergraduate students began receiving silver/nickel diplomas. Retiring faculty members, upon receiving emeritus status, are awarded silver certificates recognizing their service to Mines.

founding and funding, however, was no small matter, particularly in a depression economy. With an initial gift of $1,000 from the Franco Wyoming Oil Company, initiated by W. D. Waltman '99, the foundation was off and, if not running, at least progressing toward a financially brighter day. Given the Depression and the war to follow, however, the foundation saw relatively little activity until about 1946. Nevertheless, despite its somewhat modest start, the foundation in its seventy-six years of existence has been, and will continue to be, a vital cog in the overall educational and research efforts of Mines, probably to a degree that even the energetic and optimistic founders dared not imagine.

As the Depression became more intense, Coolbaugh, in an insightful and humorous moment, commented that "it has become fashionable to be broke, whether you are or not." Mines was not broke; but, like many other institutions of higher education, it was financially bent, with student enrollment obviously of concern. Statistics during the 1930s indicated dramatically that education, especially that of college or university, was a key to escaping the depths of the Depression. For many people of college age, however, higher education was financially beyond their reach.

Enrollment at Mines in 1932–33 was 601 (186 freshmen, 167 sophomores, 134 juniors, 85 seniors, and 29 graduate students). By 1934–35, it had dropped to a low of 496 (196 freshmen, 114 sophomores, 96 juniors, 70 seniors, and 20 graduate students). Fortunately, the slight increase

A relaxed baseball team—1930.

in freshmen helped to alleviate the drop among sophomores, juniors, and seniors. Enrollment began to rise in 1935–36, reaching a 1939–40 record of 829 (251 freshmen, 232 sophomores, 203 juniors, 127 seniors, and 16 graduate students) before beginning another decline in 1940–41 because of World War II. Thus, excepting the middle of the decade, Mines was able to maintain a relatively respectable enrollment through the Depression. It was, moreover, able to attract students from wide areas of the United States and from numerous foreign countries to keep its cosmopolitan atmosphere and its national and international reputation intact. In 1938, for example, the enrollment consisted of 52.7 percent from Colorado, 40.2 percent from other states, and 7.1 percent from foreign countries.

To keep a tighter rein on the purchase of equipment and supplies, Coolbaugh abolished departmental budgets. "Abolishing the budget," according to his view, "simply means that the various departments will not be allotted any specific sum of money for the conduct of their activities. Heretofore, budgets have been submitted for the approval of the Board of Trustees and the State Auditing Board. These budgets have been used for determining operation expenditures" (MM March 1932). Although Coolbaugh recognized that the school was not in a financial crisis, he knew that keeping it healthy would necessitate the lowest possible expenses. In another cost-saving move, the Board of Trustees reduced the 1932–33 salaries of the faculty, using a sliding scale from one percent for the lower salaried individuals to eight percent for those in the higher brackets. Coolbaugh, however, expressed optimism in a New Year's letter (1933) to the members of the Alumni Association:

> I enter the New Year with confidence and without fear. I believe that all of us as individuals, and collectively as an institution, are prepared to weather the remainder of this economic storm. The School will have to curtail expenses wherever possible, and no development or expansion can be considered until better times return. We shall maintain our present high standard of instruction. That will be our principal aim. (MM January 1933)

Beyond daily operational expenses, the school needed additional and updated instructional space to keep up with the advances in the mineral industries. The biennial reports for both 1928–30 and 1930–32 were essentially identical in the Needs and Recommendations sections. New courses introduced were Petroleum Engineering, Fuel Engineering, Ceramic Engineering, and Geophysics. Not only were new buildings required, especially for Petroleum and Geology, but also older ones needed

The Mines rifle team—1930.

Ready for Senior Day—1936.

considerable remodeling to make them more useful for instruction as well as for protection from fire. Estimates for a new building for Petroleum Engineering or Geology or for remodeling older buildings came to $200,000. The Twenty-eighth General Assembly provided for an appropriation of $75,000 for remodeling the Chemistry and Physics building; but because of obvious financial constraints, no money was available for the appropriation. The dawning of a better financial day for Mines was still in the future.

Coolbaugh, at this time, was a member of the Executive Committee of the Division of Mineral Industry Engineering Education, a branch of the American Institute of Mining Engineering (AIME). At the 141st meeting of AIME in New York in February 1932, he heard a report made by the Founders Society, a group of leading technical organizations, indicating that the employment situation of engineers in the mineral industries was considerably better than that of other fields of engineering. Again, not exactly the dawning of a new day, since all fields of engineering were being downsized, but this was at least a bit of good news for both Coolbaugh and Mines alumni.

In comparing the mineral industries with agriculture, an editorial in MM, quoting congressional figures, noted the significant difference between the federal appropriations for the two industries. Perhaps because of the larger block of voters representing agriculture, that industry received appropriations of over $109 million, while the mineral industries received only $2.811 million. Yet, mineral industries, having 5,523,454 paid workers contrasted to 13,617,811 paid workers in agriculture, furnished 74 percent of the freight for the railroads and paid approximately twice as much income tax as the agricultural industries. The editorial concluded from these figures that for every dollar appropriated for agriculture, $1.56 was paid in Federal income tax; yet for every dollar appropriated for the mineral industries, $118.21 was paid in Federal income tax (March 1932). Thus, agriculture received back 63.4 cents of every dollar paid in Federal income taxes, while the mineral industries received in return a mere 8.5 mills (March 1932).

During this period, according to Irving A. Palmer, the mineral industry of the states of Arizona, Colorado, Nevada, Utah, and Wyoming sustained

A Cabinet of Minerals Becomes the Mines Geological Museum

The *Jarvis Hall Monthly* of January 1871 notes that following the completion of the school's building, one important item was still needed: "We refer to a cabinet of minerals. This should be by all means a very complete one, that shall fully represent the rich and varied mines of our incomparable Territory." With this in mind, the trustees appealed to mine owners and others to furnish specimens for the cabinet and to "our brethren of the press" to give it as much publicity as possible. With this humble beginning, the cabinet of minerals became one of the world's fine geological museums.

Formally established in 1940 with J. Harlan Johnson as curator, the Mines Geological Museum opened its doors in March of that year in the newly constructed Berthoud Hall. Under Johnson's care, the museum grew steadily through purchases and donations of specimens. When Johnson retired in 1957, however, the position of curator was not filled. When the school received the Colorado State collection of 11,177 mineral specimens in 1965, Jerri Hamilton was hired as curator. The curator's position has since been permanently filled, and the specimens acquired by the museum have increased steadily.

The functions of the museum as laid out in 1942 specified the following:

1. To collect material and to care for material presented to the school.

2. To exhibit material.

3. To build up research collections and to stimulate research by providing a basis for graduate work.

4. To actively assist in the teaching work of the school by presenting exhibits to illustrate the various courses, by developing research study collections, and by the personal activity of the curator in supervising graduate work based on material in the museum collections.

5. To publish and otherwise disseminate information to the public.

In carrying out these functions, the museum has become an important part of the Mines academic program, an enrichment experience for busloads of public school students, and an attraction for tourists and visiting scholars. Jarvis Hall's original cabinet of minerals more than met the expectations held for it.

In the fall of 2003, the museum was moved into the newly constructed Hall of Research, a move that will enable it to meet the growth expected in the twenty-first century. For a comprehensive view of the museum's growth through the mid-1980s, see "The Geological Museum, Colorado School of Mines," *The Mineralogical Record*, May–June 1985.

its strong position, both nationally and internationally, and was producing large quantities of gold, silver, copper, zinc, and lead, as well as coal. Natural gas and molybdenum were also gaining in importance. The United States in 1930 produced almost 25 percent of the world's silver, ranking second to Mexico. The handwriting, however, was on the ever-present wall, at least for gold and silver. Palmer noted, for example, that Colorado's annual gold production decreased over a period of twenty-five years from $28 million to $4.5 million. Silver, moreover, had declined from $25.84 million in 1893 to $4.435 million in 1930. Comparable output declines were evident in Nevada. Because the production of both gold and silver was now coming as a byproduct of the mining and metallurgy of copper, lead, and zinc, the production of gold and silver, according to Palmer, would increasingly be subject to economic conditions governing the mining of copper, lead, and zinc ("The Mineral Industry in the Mountain States," MM March 1932).

Palmer's article is significant for the picture it provides of the mineral industry at the time, as well as for future implications. If the economic forces that brought on the Depression were beyond human control, so too would be those in the second half of the century, as they impacted greatly the mineral industries in America and throughout the world. As a collegiate institution, Mines was about to begin a redefinition of its educational mission that would continue to the present day.

II

That "the old order changeth, giving way to the new" is axiomatic. Everything in the history of Mines thus far has reflected a compromise between the old and the new. Presidents, faculties, trustees, students, curricula, and traditions all change—though they changed more slowly during the Depression. Coolbaugh understood that change for change's sake could easily be foolhardy, yet he knew that Mines had to prepare to meet new challenges and to reach new goals. By the 1930s, Mines' mission was still focused on the mineral industries, with degrees in Mining, Metallurgy, Petroleum, and Geological Engineering. Dean Jesse R. Morgan emphasized in "The Changing Curriculum" that the goal of Mines was "to train world engineers and not specialized tradesmen." He continued: "A man becomes a successful specialist only after he has had a thorough basic training in fundamentals and at the same time learns to think in terms of broad practical

Seniors take to the streets on Senior Day—1936.

Blue Key—1936.

accomplishment and not in a narrow specialized field" (MM July 1933).

Increased entrance requirements meant that the faculty could expect more dedication from their students and get it. Both Physics and Chemistry were added to the fifteen required secondary-school units, making a total of ten required and five elective. Courses changed from year to year, according to Morgan, because the school needed to keep pace with changing conditions in the mineral industries. Newer adjustments included eight hours of Economics in the junior and senior years, English Composition in the freshman year, Types of Literature in the sophomore year, Report Writing in the junior year, and Engineering English in the senior year. Alumnus Ben Slothower '50, in a letter of 3 December 2002 to current president John Trefny, commented on the addition of English Composition to the freshman-year curriculum. His uncle Benjamin H. Hill served as a Mines trustee in the 1920s and 1930s and was also an independent geologist and mine lessee in the Cripple Creek area. Hill received numerous letters from Mines students looking for summer jobs and from graduates seeking permanent employment. "And he was appalled," writes Slothower, "by the degree of illiteracy exhibited in these letters. So he took the issue to President Coolbaugh and together they promoted the very first required course in English composition at Mines." Uncle Ben thereafter reported remarkable results from this immersion in English composition. The Petroleum Department also revised its curriculum; and new courses were added in the Mining, Metallurgy, and Geology Departments. Moreover, such basic engineering subjects as hydraulics, physical chemistry, and thermodynamics were transferred from electives to requirements in all options.

Early in his administration, Coolbaugh and the Board of Trustees had recognized the great need for geophysicists throughout the world. It had been over sixty years since Col. E. L. Drake had started the oil business near Titusville, Pennsylvania, in 1859 with a well sixty-nine feet deep. By the 1930s, wells at least two miles or more deep would be required in some places to reach the so-called "black gold." The study of petroleum at Mines worked its way into the curriculum through geology and chemistry and evolved into the Petroleum Engineering Department in 1922. By the 1930s, the department was well on its way to an international reputation, with graduates filling positions at large oil companies throughout the United States and the world. By this time, finding oil required the skills not only of geologists and petroleum engineers, but of a new breed of geologist, as noted by C. H. C. Braden: "And when geologists can't find any clews [sic] on top of the ground, and it costs too much to dig down for evidences of oil, they put their past knowledge together with a crew of geophysicists. These geophysicists, with a stick of dynamite in one hand and a table of logarithms in the other, work

Glee Club of 1935.

Chapter 5: Reaching Out

Vacation with a Switch
(from Jack Pardee '36)

Freshmen Ted Lawson, Martin Brown, Clyde Carson, and I piled into the Model A Ford and made the trip to Estes Park without incident during the semester break in January 1933 for a few days of winter fun. The weather was clear and brisk with a foot of snow on the ground. Our destination was the vacation cottage owned by family friends of one of our group. Inside, our first concern was to get a crackling fire going in the fireplace. Immediately, smoke billowed out into the room. We rushed outside and found a ladder. One of us scrambled up on the roof and removed the flat stone from the chimney top, thus un-winterizing the place.

We aired out the room, threw another log on the fire, and our existence soon became comfortably warm. So, it was time to warm up our interiors. Now, in planning for this trip, Ted, with admirable foresight, had somehow procured from the brewery a pint of "grain," as ethyl alcohol was known in those days. Ted broke out the bottle and a juniper mix. We clinked our glasses but soon discovered that something was radically wrong. We tasted the juniper but not the stimulant. A spoonful of the liquid from the bottle snuffed out a lighted match. Our suspicions were confirmed. The stuff was plain water.

The mystery was solved by simple deduction. In Golden, Ted roomed in a house owned by a minister and his wife. This good gentleman was opposed to the possession and consumption of alcoholic beverages for social purposes. So at a time when Ted was attending classes, the bottle of spirits was secretly removed from his closet shelf. No doubt the "grain" went down the drain, and the bottle was refilled at the tap. The switch was made. So we laughed it off and otherwise had a good time with card games, bull sessions, and snow play.

Ted took exception to the deception, however, and soon after our return to town made a switch of residence address.

---

their seismographs and other gadgets and tell the geologists how the rocks lie" (MM February 1936). A clear and comprehensive view of the oil business at this time is *This Fascinating Oil Business* (1940) by Max Ball '06.

As noted earlier, Dr. C. A. Heiland, the noted German geologist and geophysicist, began teaching courses in geophysics in 1926. By the 1930s, four undergraduate courses and seven graduate courses were being offered as an option in Geology, an option that was gaining international recognition. Heiland explained how Mines was helping to meet the growing demand for geophysicists:

Anticipating the future needs of the industry and keener competition among geophysical

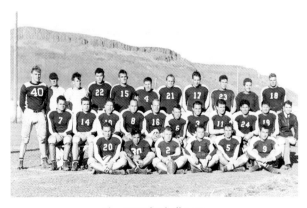

The 1936 football team.

Selected members of the class of 1936.

engineers, the Colorado School of Mines has been almost unique in providing a training that gives the geophysicist not only his geological background but his requisites in mathematics and physics as well. As a consequence, the number of undergraduate students has grown steadily, so that almost 200 students have received undergraduate geophysical training during the past ten years. As far as we are aware, there is no one of these who is unemployed at present. It speaks well for the type of training offered that we are able to place our students during the depression. (MM September 1936)

In addition, Mines turned out seventy students with graduate training in geophysics between 1926 and 1936. By 1936, because of the Depression, the department's faculty had been reduced from six to three. Heiland believed that the department needed better facilities and a larger faculty if it was to maintain its international reputation. A colorful and dynamic personality, Heiland proved a fortunate choice to bring a new department to life. Frank Hadsell (Professor Emeritus of Geophysics), in his unpublished history of Mines provides an insightful description:

> Putting Carl Heiland at CSM in 1926 was much like running an Intel Pentium in your old IBM PC, or to use a more contemporary analogy putting the engine of a Stutz Bearcat in your Model-T. Heiland had very little tolerance of ignorance and even less of indolence. All his students recall, still with some trepidation "und des obviously follows from dat" and ten-hour open-book examinations that decided your future, not only at CSM. This explosive situation was kept under control largely by Dr. F. M. Van Tuyl, the Head of the Geology Department and the new president of CSM, Melville F. Coolbaugh. It was not until 1949 that Geophysics became a full-fledged department at Mines.

Another dynamic of a Mines education during this period was the summer school program. Although Mines had offered summer courses as early as 1912, a formal summer program was begun in 1923. From its inception, this summer school program grew slowly but surely. Dean Jesse Morgan, who directed the program, observed that because Mines was the only school in the country offering engineering training specifically for the mineral industry, its specialization would be carried over into the summer school through regular-term courses taught by members of the regular faculty (MM July 1934). In the year 1937, three hundred students enrolled in the summer school, representing twenty-seven universities, both domestic and foreign—most of whom were seeking advanced standing at Mines or making up deficiencies at their own schools.

Class of 1936 fiftieth reunion—looking good.

Class of 1936 sixtieth reunion—still looking good.

Professor William B. Plank of the Department of Engineering at Lafayette College studied forty-eight schools in the United States and six in Canada granting degrees in mineral technology for the 1934–35 academic year. The study illustrated dramatically the significant niche Mines was filling in the production of mineral engineers. Of the 5,011 students in the study, 4,624 were undergraduates, and 285 were graduate students. Of these, Mines had 537 of the undergraduates (11.6 percent) and 11 of the graduate students (3.8 percent). Of the 727 total graduates in the spring of 1935, Mines had 87 (12 percent). Of 224 receiving Mining Engineer degrees, Mines had 35 (15.6 percent). Of 174 receiving Metallurgical Engineering degrees, Mines had 27 (15.5 percent). Of 177 in Petroleum and Natural Gas, Mines had 15 (14.1 percent). Of the 77 Geologists, Mines had 9 (11.7 percent). Among the 54 schools, Mines was well above average in producing graduates for the mineral industries. (See *Mining and Metallurgy* November 1936 for a complete report of Plank's study.)

III

While the curriculum was evolving during the early 1930s, student life and traditions generally followed past patterns, particularly at the beginning of the school year. Freshmen were welcomed and instructed in Mines traditions by President Coolbaugh, student-body leaders, faculty members and coaches, and alumni. The sophomores won the Tug-of-War across Clear Creek against the freshmen in 1932, though they were accused of utilizing a two-ton truck in the process. The sophomores also won the Barbecue, during which freshmen faces were blackened in the Assay Laboratory. The fraternities gave up three years of deferred-pledging experimentation and returned to an anything-goes "cut-throat" style of rushing, the goal of which was to let the freshmen appraise the fraternities and vice-versa. Because there were no school dormitories, it was necessary for the freshmen to secure accommodations before classes began and for fraternities to reach their quota of pledges to maintain their financial positions.

A new tradition of granting silver diplomas began with the class of 1934. The five-by-six-inch diplomas were etched upon a sterling silver plaque and enclosed in a leather case embossed with the school emblem on the cover. Gaylord Warren of the graduating class originated the idea that was heartily accepted by the administration, faculty, and

The *Oredigger*

Campus newspapers have been one of the more significant extracurricular activities in American colleges and universities as purveyors of news, forums for debate, and crusaders for causes. In those functions, they are much like proprietary newspapers. The quality of campus newspapers varies considerably, depending upon such variables as size of the school, the presence of journalism schools or courses, and pay for editorial and business staff. The presence of a newspaper, regardless of its quality, is important to the campus scene.

Mines' *Oredigger* was first issued on 21 March 1921 as a substitute for the *Mines Magazine*, which ceased publication for a while when the alumni association broke from the school. From the many names suggested, the title proposed by A. K. Seeman '22 was accepted. Seeman is also credited for writing several verses of the "Mining Engineer." The first editor was Ronald K. DeFord '21, who successfully guided the paper through its initial year. Starting as a tabloid-sized paper, it was eagerly accepted by students, alumni, and friends of the school. The goal of the paper has been to present campus news of all kinds, to provide students with a forum, and to campaign for improvements for the school. One of its first endeavors was support for the construction of Brooks Field. The *Oredigger* has maintained a comparatively high level of journalistic quality and coverage, as reflected in positive ratings from the Associated Collegiate Press' critical service. It has not shied away from controversial topics in its coverage, and it has attempted to maintain a balance of viewpoints. Like all newspapers, it has had its critics from time to time—students, faculty, administration, alumni, and friends.

Contemporary times have not been overly kind to campus newspapers. Many are experiencing financial problems and lack of interest among students to work on or even to read the paper. The *Oredigger* is not immune from such problems, as indicated by the following excerpt from a farewell editorial from Bryce Lakamp that appeared on 3 December 2003:

> The *Oredigger* has been going through times of constant change in recent years. Two main reasons are the lack of student involvement and the lack of necessary funds to even produce the paper. The *Oredigger* writing staff started this semester with 30 individuals. Now there is a staff of 5.

> Within the past three semesters, the length of the *Oredigger* has been cut in half. Two years ago the average issue was 16 pages. Today the average is 8. The *Oredigger* as a club is starting to dwindle. If the current lack of involvement continues, it is possible there will not be a paper next year.

Along with the disappearance of pay for the editorial staff was the class credit that until recently was available through the Liberal Arts and International Studies Division. Change at Mines, as at any school, comes not just in curriculum, faculty, or administration. It is pervasive.

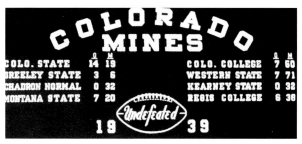

The scoreboard tells the story.

students. Students from the Engineering Drawing Department, under the direction of Professor George W. Salzer '21, did the design, including the hand-lettering. Carl Blaurock '16 provided and rolled the silver, and the school's instrument designer, Charles Hill, supervised the etching. Fitting the wording to such a small diploma fell to the administration and to Max Ball '06. The silver diploma was customary for all graduates through 1969, after which time it was given only to those receiving graduate degrees. While silver diplomas glossed the 1934 commencement, former U.S. President Herbert Hoover burnished that of 1935 when he was granted an honorary Doctor of Engineering degree before a standing-room-only crowd.

The Mines spirit was thus still alive and thriving in all respects even during a lack of success on the gridiron. As today, Mines' first object in the 1930s was to train engineers who would bring note to themselves and to their alma mater as they followed their professions wherever they might lead. Perhaps because there has never been any favoritism for athletes at Mines, the won-lost records have often been disappointing. The academic stature of the school, however, has been maintained to the fullest degree.

As noted previously, football early on was an important sport at Mines. And so it was in 1930. Most players on that team were, of course, from Colorado; but other Western states, as well as Illinois, Indiana, New Jersey, Ohio, Pennsylvania, and Virginia were represented. The aggregate weight of the forty-two players was 3.4 tons, with an average per player of 162 pounds. The lightest player was a halfback who tipped the scales at 117; the heaviest was a tackle at 197. The shortest player was just 5 feet 4 inches tall, and the tallest was 6 feet 1 1/2 inches. Only a half-dozen players reached six feet in height. This was a mean and lean team of natural muscle and bone with no iron pumping and no steroids.

The *Oredigger*'s survey of football players in December 1933 showed that nine believed that football interfered with their studies, while eleven did not. Thirteen felt that they would spend more time on studies if they were not playing football. Another thirteen felt that they were often too tired to study after practicing football. Some of those polled felt that the administration was at fault for the lack of success on the gridiron. An editorial summed up the situation:

> We must admit these conditions but as Coach Clark said recently at a meeting of the Colorado Section of the Alumni Association we must recognize the fact that no athlete becomes proficient in any sport without

The All-American Lloyd Madden.

These 1939 Miners could play both ways.

practice. Of two athletes who have the same ability the one that practices the game the most is the better player. And so it is with any two athletic teams; the team having the greatest number of experienced players is usually the winner. It occurs to us that here may be one of the reasons Mines teams have not annexed more victories in late years. Some of the men on the first team never played football before coming to Mines, yet on other teams of the conference it is noted that inexperienced players among them is a rare exception. (MM December 1933)

When George Scott came to Mines in 1934 as athletic director and football coach, he saw his new position as a distinct challenge:

> Here was an opportunity to become coach at a college which during the last decade had won very few games. Athletics seemed at a very low ebb. Many alumni of Mines and most other people informed me that the situation was hopeless. Friends implored with me that I was making a mistake which I would soon regret. Somehow all of my life I have got a lot of fun working with difficult problems. Here was a problem that seemed really difficult. But after working for two months and seeing my team lose more games than I had formerly lost in 11 years, I am pleased with the situation and the outlook. The work is more interesting than I imagined it could be. (MM November 1934)

An inveterate optimist, Scott was impressed with the cooperative and friendly spirit of the campus, as well as the lack of adverse criticism. He felt, moreover, that Mines would soon be able to compete with its opponents in the Rocky Mountain Conference. He was pleased that Coolbaugh believed firmly in the value of athletics at Mines: "In Dr. Coolbaugh's own words if he were asked to choose between two men equally qualified scholastically for a difficult position and one of them was a capable athlete, he would by all means choose the athlete." Scott's 1934 Mines football team scored a total of 24 points to its opponents' 182 in a season of one win and seven losses. The 1935 season of nine games was little better; Mines won two and lost seven, with one of its victories over Grand Junction Junior College. The basketball team of this year, not to be outdone, suffered through fourteen straight defeats.

Football of the 1930s, while exciting enough and played by some very skillful players, had not yet become the science that it is today. To be sure, Ivy League schools and large state universities played

Cheerleaders: few in number but loud in voice.

Rank Hath No Privileges

Dave Coolbaugh '43 (son of Mines President Melville Coolbaugh), a student at Mines in 1941, failed two courses in the fall term but passed them in the spring. Toward the end of the spring term, Dave was notified to report to the president's office immediately. After an unexplained ten-minute wait during which time no one came out of his father's office, Dave was told to go in. Recounting for Dave the academic problem of the fall term and showing no interest in Dave's having passed them in the spring, President Coolbaugh pointed his thumb over his shoulder and informed his uneasy son that he was OUT, suspended until the 1942 fall term—noting that it was time for him "to sample the cold, cruel world." Dave was aware that his two brothers had suffered the same fate, Jack for good reason and Frank for getting married without parental blessing. With his spirit low, Dave left his father's office and began searching for a job.

With the help of some of his professors, Dave found a training position with the Pure Oil Company and joined one of their crews in Montana in August 1941. With the attack on Pearl Harbor and the American entry into World War II, Dave enlisted in the Navy. Advised to return to school and graduate so that he could receive a commission, Dave telephoned his father and "laid my cards on the table." "Come on home" was the simple reply from a father who knew his son perhaps better than the son knew himself. Dave returned home and reentered Mines, deciding that because of the seismic experiences he had just gone through, his future would be with mining geophysics rather than in the petroleum industry." Receiving his degree in May 1943, followed by his commission, Dave spent the rest of the war with the Navy in the Pacific theater.

---

before huge crowds and made headlines on sports pages across the country. Customary rivalries were carried on among collegiate institutions of all levels and sizes. The Mines-Colorado College rivalry, for example, lasted ninety years. Few communities of the time could boast elaborate playing facilities or highly trained coaching staffs at the high school and junior high school level—lest they be like Massillon, Ohio, a steel town that in the mid-1930s filled Tiger Stadium with 20,000 fans on Friday nights to watch Coach Paul Brown's perennial state champions demolish sixty-six opponents in a row. Neither was recruiting high school players the sophisticated process that it has become today as more and more institutions strive to qualify for, and win, at the National Collegiate Athletic Association's (NCAA) Division I-A level.

With John Mason as new coach and a welcome influx of talent, Mines ended the decade on a resoundingly high note of success: an undefeated football team in 1939 that boasted eight victories, making it one of the ten undefeated teams in the nation for that season. The Blasters, as Mines teams were called then, scored 268 points, an average of 33.5 per game while allowing opponents only 37 points, an average of 4.6 per game. The average margin of victory, then, was 28.9, marked by 71–7 and 50–7 routs over Western State and Colorado College, respectively. Though the Blasters may not have done so well against Texas A & M, Tennessee, or Cornell, three of the other unbeaten teams, they surely would have won more than their share from the other six smaller programs: San Jose State, Louisiana Normal, Maryville Teachers, Slippery

Beta Theta Pi house.

Rock Teachers, New Britain Teachers, and DeSales. But, speculation aside, the 1939 season provided moments to remember. Lloyd Madden led the ten major conferences of the time in scoring with 141 points, beating out Don Lieberum of Manchester and the great Tom Harmon of Michigan. Behind an offensive line of De Goes, Moe, Geiskieng, Thornton, Yeager, Katzenstein, and Flynn, Madden carried the ball ninety-two times for a total of 1,316 yards. Both Madden and Katzenstein were named to the Little All-America team. This Mines eleven was a team well manned at virtually every position on offense and defense, a team for which coaches pray but seldom see. In 1940, Mines fell to a more customary level, managing three wins against four losses.

Another area of success for Mines during the trying 1930s was the physical plant. Although many schools during the Depression were thankful just to be able to carry on daily operations, let alone constructing new buildings, Mines, with Coolbaugh's leadership, not only electrified the "M" in 1931, but also added two much-needed and valuable buildings to a campus that had accomplished no new major facilities in a number of years. The Field House, built with state money, was completed in 1937 and eventually named in 1949 in honor of Frederick C. Steinhauer '99, who served as a member of the Board of Trustees for twenty years and as president for seven. Berthoud Hall, one of the many projects carried out nationwide by the Public Works Administration and still one of the more stately buildings on campus, was completed (its middle portion) in 1939 and became the new home for the Geology Department and the museum. These buildings allowed Mines to carry out a more effective educational program and added much to the physical appearance of the campus, which was beginning to equal the school's excellent academic reputation.

IV

As the 1930s drew to a close, Japan and Germany, both countries with prevailing military traditions, were gaining military, political, and economic power. The Treaty of Versailles that marked the end of World War I reflected the desire of France and Great Britain to weaken Germany and to effect a more favorable and peaceful balance of political and economic power in Europe. Unfortunately, the treaty ensured that in just over twenty years another conflagration would sweep Europe and the world. Japan was seeking to ensure that the rising sun would continue rising and was willing to risk much to keep from being economically and politically constrained by the West. The Soviet Union, moreover, added a Marxist threat to the capitalist world that would have to be dealt with sooner or later. Hitler, Mussolini, Stalin, and Togo were ready to stir this new witch's brew into a six-year conflict of a scale that few could imagine.

While World War I did much to break the American tradition of isolation from the problems of Europe,

Sigma Nu house.

Kappa Sigma house.

there was still considerable feeling that Europe should be left to its own designs in dealing with Mussolini, Hitler, and Stalin. This time, however, America was not to have the luxury of standing on the sidelines until the conflict was almost over. If 7 December 1941 was a day to "live in infamy," it was also a day to bring both immediate and long-range change to America and the world.

Colleges and universities again felt the impact of war almost at once, particularly in terms of reduced enrollments. Instead of studying for examinations, playing football or basketball, carrying out high jinks on days like Mines' Senior Day, or looking forward to practicing trained-for professions, the young men, at least most of them, from ages eighteen through twenty-two would soon be learning and practicing the various necessities of warfare.

As noted earlier, Mines was, in 1919, one of the first schools to be granted an ROTC program. That the school and the military made a good match for twenty years was proven not only by the number of officers that Mines turned out, but also by the numerous individual and unit awards garnered by its cadets. Between 1920–21 and 1941–42, over 847 Mines cadets received commissions, an average of almost forty per year. In 1941–42, 414 students were enrolled in ROTC courses—over half the student body. Little wonder that Mines was often referred to as the "West Point of the Rockies." An estimated total of more than 2,000 alumni and former students served during the war. Mines, with its 847 commissions, led the nation's schools in the number of engineering officers trained through the ROTC.

The War Department contracted with many colleges and universities to aid in specialized training of troops, for which the schools were reimbursed. From June 1943 to April 1944, Mines quartered and trained in basic engineering a detachment that reached a peak of 539 cadets. The west wing of Berthoud Hall, utilized as a mess hall, must have resembled a miniature version of that of West Point, though in cafeteria-style. It fed the entire detachment in one hour. Butchers and cooks daily prepared large slabs of meat and cases of fresh vegetables, and bakers turned out loaves of fresh bread from a battery of ovens. Some 1,200 glasses, 600 coffee cups, 600 bowls, and 600 compartmentalized metal trays were used each day, arranged on thirty tables seating ten men each. These soldiers no doubt created their own version of "Mines spirit" in their short stay on the campus. This influx of military students helped sustain the school financially as the civilian enrollment dropped to a low point in the fall of 1944.

As a predominantly male school, Mines could not count on any influx of women students to help stabilize the enrollment. Ironically, the 121 civilian

Sigma Alpha Epsilon house.

> # The Great Shoot-Out at the Corner
> ## (from Fred Hynes '41)
>
> Two fraternity brothers, Pete Leidich '43 and Johnny Masterman (who later graduated from Penn State), were arguing about the effective range of a twelve-gauge shotgun. Pete contended that such a gun loaded with #8 shot would not wound him at one hundred feet. Johnny accepted the challenge. After efforts failed to dissuade the disputants, Pete prepared himself with heavy underwear, a heavy wool shirt, a wool sweater, a denim field jacket with plaid cotton lining, and a miner's hard hat, tipped back to protect his neck. He then proceeded to the street corner one hundred feet away and stood bravely under the light post. Johnny, meanwhile, had loaded the shotgun and looked down from a second-story window from which he would fire his shot. At Pete's shout to "fire away," Johnny took a deep breath and squeezed the trigger.
>
> BAMM!
>
> In the still of the night, the shot sounded like a cannon. As if nothing had happened, Pete sauntered back to the fraternity house. When asked if he had felt anything, he responded, "Oh, a little stinging." The shot, however, had penetrated all of his protective clothing, leaving about a dozen pellets stuck in Pete's back. Reluctantly, he agreed to visit Doc Hewlett, every Miner's friend. Lying face down on the examination table, Pete suffered as Doc plucked pellet after pellet from his back—that is, all but one. As Pete lay on the table, recovering from his ordeal, he relaxed his butt muscles. To everyone's astonishment, one lonely pellet came rolling down the crack of Pete's butt to land with a ping on the examination table.
>
> After much imploring, Doc Hewlett agreed not to tell the police about the incident. After all, it was a planned shooting—perhaps butt-threatening but certainly not life-threatening.

students, including 17 discharged veterans, who enrolled for the fall classes of 1944, preceded an upward trend that soon taxed the school's faculty and facilities. In the meantime, Coolbaugh had to stanch rumors that school closing was imminent, rumors that he had heard in person as far east as New York and Cincinnati. "There never has been the slightest idea of closing the school," he said in an interview with the *Jefferson County Republican*. "It is in excellent financial condition and there is no need of closing. These rumors are doing the future of the school no good and myself and the governing body will appreciate the cooperation, especially of Golden residents, in stopping these rumors" (undated clipping in Archive).

In addition to sending its ROTC students to war and providing engineering training to a military detachment, Mines aided the war effort by carrying out research and testing for war industries in the Experimental Plant and serving as a supervising institution for engineering schools of Colorado and Wyoming. Coolbaugh acted as regional advisor for these schools and was regional representative of the district comprising Colorado, Idaho, Montana, Utah, and Wyoming. Courses were designed with government approval to train thousands of men in special engineering skills. According to the Biennial Report of 1942–44, care was taken "that the requirements of regular four-year courses in engineering are rigidly enforced and, in fact, strengthened in compliance with the expressed intentions of the Federal Government that the emergency training be superimposed on the regular work of the School and in no way supersede it." The government urged colleges and universities to extend their normal period of instruction, thus enabling them to complete as much of their education as possible before induction into the armed services. Mines lengthened its summer school from seven weeks to eight and encouraged all freshmen, sophomores, and juniors to take advantage of summer courses. Moreover, the departments of Military Science and Physical Education combined efforts to improve the physical fitness of Mines students with calisthenics, military ranger tactics, and judo or ju jitsu.

V

Although war was a grim reality for a great part of the world in the spring of 1942, American college students were for the most part still shielded from that reality. On 16 April of that year at 8:05 a.m., however, a dynamite blast on the Mines campus defied Coolbaugh's decree that Senior Day be abolished. This particular day of its own infamy once again occurred amidst alcoholic exuberance, outlandish costumes, and high jinks, topped off with the traditional senior newspaper. This time the paper was titled *The Senior Challenge*, a title

Sigma Phi Epsilon house.

ATO house.

considerably less colorful and shocking than usual for the once-a-year publication. Moreover, the seniors promised, among other things, not to use profanity or personal abuse in their paper or to disrupt Golden public schools. They went so far as to establish a fund to ensure payment for any damage done and to provide a police force of twelve seniors to ensure a more subdued Senior Day. The paper, on the other hand, urged all seniors to stick together and "act as a unit to counteract any unjust reprisals by the faculty or administration." The gist of the paper, perhaps both jocular and serious, was that Mines, in the view of the more vociferous and self-anointed seniors, needed a drastic overhauling if it were to remain the leading school in providing engineers for the mineral industries.

Criticism ran the gamut from faculty and administration to courses offered and degrees granted. A copy of the paper, with a letter explaining its purpose, was sent to the Board of Trustees. How much basis there was for the complaints mentioned is difficult to assess. Certainly some of the complaints were those of young men about to say goodbye to several years of hard study and who were taking one last opportunity to vent their frustration against a system that not only educated them, but in the process put them through an academic "gauntlet" that generally required more than four years to complete.

The "ghosts" of Tau Beta Pi drink a toast to new members.

Aware of the problems of Senior Day over the years and concerned about the damage done in a near riot the previous year, Coolbaugh was ready to put the quietus on any more such celebrations. While most seniors wanted Senior Day to be continued, a fair number thought that in light of the wartime situation, it should not be. In Washington on defense business on 16 April, Coolbaugh wasted no time in taking action when he returned, dismissing two seniors for insubordination and withholding the diplomas of seventy-five more for participating in Senior Day activities.

Here, indeed, was a clash between a cherished tradition and disobedience of an administrative order. It was a no-win situation for either side that provoked numerous newspaper articles and personal letters to Coolbaugh. An alumnus in Bartlesville, Oklahoma, wrote, "The feeling of the Bartlesville alumni is practically unanimous in backing you up. The question of Senior Day will, of course, always be a debatable issue, but there is no doubt in anyone's mind that you were correct in enforcing discipline." A handwritten postcard from a Denver woman stated bluntly, "Stand pat & do not let them graduate at all. Who wants to hire a mature senior guilty of insubordination? Uncle Sam frowns on that & they should be punished in a way they will remember all their lives." Many letters from parents of the seventy-five students whose diplomas were withheld accused Coolbaugh of being insensitive to the fact that their offspring would soon be off to war. Coolbaugh responded to each of these letters, stating "there is no thought that this disciplinary action will jeopardize your son's career or that it will prevent him from ultimately receiving his degree. The fact of the matter is that he will have learned the lesson that insubordination cannot be tolerated in any organization whether it be educational, governmental, or private." An editorial in the Cripple Creek *Times-Record* of 27 April 1942 does not question Coolbaugh's disciplinary action but criticizes the decisions to withhold diplomas and to abolish Senior Day, particularly because Senior Day touches on the meaning of the Mines spirit: "If these seniors are not allowed to graduate, the school and students will receive publicity which is harmful to both. The Miners [Mines students] are not rough; they are tough. They are just the way we want them to be. We are going to need a lot of tough men in the next few years." The *Transcript* of 30 April 1942 echoed such sentiments:

> In the arsenal of defense and on the front, Mines men have proven to the army, navy, and air corps that Mines turns out reliable, trustworthy and capable graduates, so much

Berthoud Hall—a classic example then and now of WPA architecture.

Chapter 5: Reaching Out

ROTC cadets—soon to go to war.

so that the institution has been given special deferment and a preferential rating by the government. That Mines men accept responsibility, respect authority and make good is a school tradition worth maintaining at any cost. It is the oldest tradition of the school, starting with the first graduating class. It gives the school a standing in its special field on a par with the United States Military Academy at West Point.

And so Mines survived another Senior Day. The seventy-five students did eventually receive their diplomas, and many did go to war. Senior Day, moreover, was not permanently abolished. And Mines' problems with this traditional day were not over—not by a dynamite's blast.

The Allied victory over the Axis powers was finally achieved in 1945, bringing with it the promised dawning of a new day both for the world and for the Colorado School of Mines.

CHAPTER 6

A Decade of Change

If we are to bring the broad masses of people of every land to the table of abundance, it can only be by the tireless improvement of all our means of technological production and by the diffusion of every form of education of an improved quality to scores of millions of men and women.
—Winston Churchill

The war in Europe ended in May 1945 and in the Pacific the following August. If World War I brought the end of three European monarchies and the creation of many new nations, World War II gave birth to two superpowers and signaled the end for two major European overseas empires. The first half of the twentieth century bloodied the world almost beyond human imagination. Millions upon millions of soldiers and civilians were victims of industrialized killing, and numerous cities were turned into rubble. But with peace, the healing could begin. Those who gave their lives in the war were mourned and honored. Songs like "Comin' in on a Wing and a Prayer," "Don't Sit Under the Apple Tree with Anyone Else but Me," "Let's Remember Pearl Harbor," and "The White Cliffs of Dover" faded along with memories of rationing—coffee, gas, meat, shoes, silk stockings, sugar, tires.

In America, at least, the beginning of the second half of the twentieth century looked promising, even as allies became enemies, and enemies became allies. The troops came home, welcomed by the GI Bill passed to ameliorate unemployment and to provide opportunities for higher education. The boom of bombs gave way to the boom of economics and babies. The United States and the Soviet Union were about to begin a cold war *pas de deux* with nuclear weapons as props, but Americans had other things on their minds.

The 1900s to this point had been marked by ups and downs: the excitement and promise of a new century, World War I, the Roaring Twenties, the Depression, and World War II. Having emerged from the last with its homeland physically untouched, its economy the strongest in the world, and its optimism and self-confidence renewed, the

Opposite page: Mines received national publicity when the Life magazine cover of 24 November 1952 presented a striking photograph of the school's student body of nine hundred men flowing out from Guggenheim Hall—with a single freshman woman at the focal point. At the time, Nancy Easley was the lone coed on the Mines campus. (Time, Inc.)

nation, even with the hovering threat of nuclear war, was primed for the good life of the 1950s and the leadership of a war-hero, father-figure president who resembled Mister Clean. Colleges and universities, too, were ready for calm and growth that would allow them to resume their primary function of educating America's young. The wartime problem of few students quickly gave way to skyrocketing enrollments produced by the deluge of veterans taking advantage of an education under the GI Bill. Institutions of every size and kind enthusiastically welcomed this growth and the challenges it brought.

I

The departure of Melville Coolbaugh from the presidency of Mines coincided with the end of the war. On the advice of physicians, he had asked in September 1945 to be relieved from his presidential duties, and he submitted a formal resignation in October. Providing sorely needed leadership,

Dr. Ben Parker, the ninth president of Mines.

Coolbaugh in twenty-one years as president guided the school out of a period of campus unrest and through the difficulties of the Depression and World War II to new levels of educational excellence. A leader with a definite vision for Mines, whose aspirations no doubt soared beyond what reality would allow, Coolbaugh could leave his presidency confident that the school was ready and able for future growth as it entered the second half of the twentieth century.

As Coolbaugh was completing his tenure, the Mines Alumni Association, on 29 and 30 September and 1 October 1945, celebrated its Diamond Jubilee with 1,650 in attendance, including over 300 in the academic procession. Several persons received honorary degrees, and a record number of alumni were given Mines Medals for their services and professional accomplishments. Present at the celebration were Robert Stockton '95, one of the founders of the Alumni Association, and C. Lorimer Collburn '07. In remarks at the banquet, they traced the development of the association through its first fifty years, stressing that its original goals of cultivating friendship and acquaintance, mutual aid, and elevation of the reputation of the school were still the pillars of the organization. Over the years, the association had expanded its efforts on behalf of the school and its alumni. Following establishment of the Colorado School of Mines Foundation in 1928, the association organized the Alumni Council to coordinate activities of the trustees, faculty, alumni, industry, and students in promoting the foundation. With the expansion of the Life Membership Fund into the Alumni Endowment Fund, the association was better able to support itself and to promote the foundation. In 1940, Frank C. Bowman '02 became the association's first executive director. Particular praise was given at the Diamond Jubilee to the thirty-five years of publication of *Mines Magazine*, a publication whose growth, influence, and prestige were instrumental in projecting the school's reputation throughout the world.

Dr. Ben H. Parker succeeded Coolbaugh in January 1946. He was the first Mines alumnus to serve as its president. Indeed, Parker received all his degrees from Mines: E.M. '24; M.Sc. '32; and D.Sc. '34. A petroleum geologist, he had extensive industrial experience before joining the Mines Geology Department in 1933. One of his first tasks was to arrange for facilities and faculty to take care of the friendly invasion of the many veterans seeking a Mines education.

Another task was to make sure the increased enrollment did not militate against the quality of that education. By 1946, as many as 1,500 applications for enrollment were submitted in a single month. Because so many applicants qualified, a priority system was necessary to ensure fairness to all: Priority 1, former and present Mines students in good standing; Priority 2, veterans who were Colorado residents; Priority 3, nonveterans who were Colorado residents; Priority 4, veterans who were not Colorado residents; and Priority 5, remaining nonresidents. Under this rather complex system, all qualified Colorado residents were admitted. As Parker recognized in his 1946–48 Biennial Report, only through holding to a carefully selected enrollment of no more than 1,200 could Mines be confident that its high standards would be maintained.

Complicating the enrollment situation was the large number of married veterans—about 30 percent of enrollees—who needed family housing. All housing, whether for married couples or single students, was extremely tight in the Golden area. Mines, moreover, would not have a permanent dormitory until 1954 (Bradford Hall). Like other schools, it turned to the government and any other available sources for help in solving the housing crisis. A mining engineer from Leadville anonymously donated four twenty-by-sixty-foot prefabricated houses that were once used to house German prisoners of war. They were sectionalized and moved to the campus and re-erected as two- and

Temporary residential units were a partial answer to the housing shortage faced by Mines following World War II.

three-bedroom apartments. The Federal Housing Administration provided thirty-five trailers, along with other unused military and government facilities that were converted into eighty apartments. The trailers were parked in Steinhaur Field House, where water and electricity were available as well as warmth. In the winter, the doors to the field house were kept closed, accentuating the various smells of cooking, inadequate plumbing, and poor ventilation. This gypsy-like living engendered a sense of identity and unity and even folklore among these young families that lasted well beyond graduation and contributed in its own way to the Mines spirit.

Housing was but a first step in handling the increased enrollment. Faculty, having been reduced drastically during the war, and classrooms were also in short supply. Because of the many opportunities in industry for engineers, Mines faced the perennial problem of finding qualified faculty members. From

Colonel Wendell W. Fertig

Through its time-honored ROTC program, Mines has produced more than its share of Army Engineer officers who have served their country with distinction. Among the many capable military officers who have commanded the Mines ROTC unit, Colonel Wendell Fertig stands high. Born in La Junta, Colorado, on December 16, 1900, Fertig attended both the University of Colorado and Mines though he graduated from neither institution. Upon leaving Mines in 1924, he accepted a commission as second lieutenant in the Army Reserve. Working for numerous mining companies in the United States and the Philippine Islands, Fertig advanced to the rank of lieutenant colonel.

With America's entrance into World War II, he went on active duty with the Corps of Engineers to fortify the Philippines against an attack by the Japanese. When American forces in the islands finally surrendered, Fertig was deep in the jungle of Mindanao. Choosing to fight rather than surrender, he and a Navy petty officer and an Army captain organized a guerrilla force of Filipinos and a small group of Americans. By 1943, with Fertig in command, this force was carrying out constant raids against the Japanese, as well as delivering strategic information to General Douglas MacArthur's forces as they prepared to return to the islands. With the defeat of the Japanese in the Philippines, Fertig was promoted to colonel and assigned to MacArthur's general staff to advise on guerrilla warfare tactics. Among the half dozen awards he received was the Distinguished Service Cross. His achievements have also been treated in fiction and film.

In 1947, Fertig returned to Mines as professor of military science and commander of the ROTC unit. At his retirement from the Army in 1950, he received an honorary Doctor of Engineering degree from Mines. For the next ten years, Fertig worked as a mining engineer and consultant in Colorado and was active in lecturing on guerrilla and psychological warfare. From 1960 to his death in 1975, he served as executive secretary of the Mines Alumni Foundation and was also honored with an honorary membership in the Mines Alumni Association. A mining engineer, war hero, and teacher, Wendell Fertig's life was indelibly marked by accomplishment.

Pictured above: Colonel Wendell Fertig, who served the nation gallantly during World War II, also served Mines as ROTC commandant and as executive director of the alumni association.

---

the end of the war until the early 1970s, competition for faculty in virtually all disciplines was intense, with salaries having to be adjusted accordingly. The overall condition of the academic profession concerned not only institutions, but also present and future faculty members. The American Association of University Professors, a watchdog of academic freedom, had become a proponent for the economic welfare of faculty. The image of the absent-minded professor living as a recluse never was quite the norm, and it had by now become even less so.

From the beginnings of American higher education, faculty members generally could not support a family of more than one child and often had to supplement their incomes through tutoring, teaching summer school, or, especially in the case of scientists and engineers, consulting. Academic life prior to World War II was marked by commitment to teaching and loyalty to an institution that in many cases overshadowed monetary concern. The new breed of faculty, however, was more concerned with material reward and individual advancement. In many institutions, research responsibilities were added to teaching loads that today's typical faculty member would consider outrageous.

The next several decades would bring many changes in faculty-administration relations. But for the time being, Mines competed in the faculty job market against other colleges and universities. Parker defined the situation in his 1944–46 Biennial

Not much, but it was home.

Report: "It seems just that a general increase [in salary] of a size sufficient to meet the increased cost of living should be given to the faculty members who have remained with the school in spite of offers of much larger salaries from industry and from other institutions," and "it cannot be expected that they will continue to make such sacrifices permanently." Parker also predicted that the upward trend in salaries would be most definite in the engineering and scientific fields.

While faculty needs were considered to be more important than facilities and could more readily be addressed, Parker was quick to emphasize the importance of a campus building program to provide classroom and laboratory space for the expected continuing increase in enrollment. As it had to do for student housing, Mines would have to rely on what were anticipated to be temporary academic buildings of a military style. Some, obtained from Buckley Air Field, were erected on the school drill grounds and provided classrooms, laboratories, office space, and a cafeteria. The first phase of the permanent building plan Parker envisioned included a new heating plant, a petroleum and chemistry building, and a physics building—with a new library, a physical education plant, an assembly building, a student union, and a

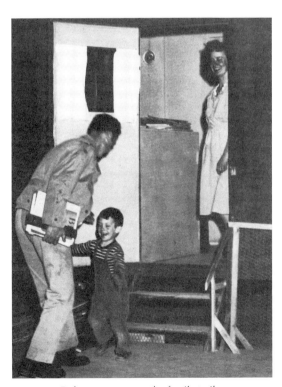

Father meets son at the family trailer.

Studying could demand no privacy.

dormitory to follow over a ten-year period. Because of the usual restrictions on funding, it would be awhile before the bulldozers would roll in earnest.

To strengthen the math background of entering students, Mines added plane trigonometry in 1946 to the basic entrance requirements of chemistry, physics, advanced algebra, and solid geometry. Also, departments continued thorough reviews of their respective curricula that began in Coolbaugh's last year, but had not been carried out since before the war. In one of his last statements as president, Coolbaugh assured both students and industry that "the content of the courses in every department would be subject to constant evolution to meet best the requirements of the times, always keeping in mind that the engineer must have an appreciation of the humanities as well as a comprehensive knowledge of the sciences" (MM August 1945). In his first Biennial Report, Parker underscored the intense faculty study of the curriculum to keep it abreast of the many changes occurring in the mineral industry: "The mining engineer, the geological and geophysical engineer, the petroleum engineer, and the metallurgical engineer all find themselves in a maze of new developments."

The Mining Department planned an increased emphasis on mining low-grade ores to counter the exhaustion of higher grades and to develop more efficient mining methods. A specialized curriculum in coal mining was added to meet a growing demand in Colorado. Metallurgy added new courses to deal with advances in physical metallurgy, nonferrous metals, and the use of alloys. Geology planned to continue emphasizing the needs of industry in its advanced courses and to emphasize exploration and new areas such as aerial geology. Another goal was to define more clearly the relationship of Geology and Geophysics as Geophysics was prepared to offer its own degree in 1949. Because of rapid developments in both production and refining areas, Petroleum was divided into two departments: Petroleum Production Engineering and Petroleum Refining Engineering, with Production Engineering more concerned with deeper drilling and Refining Engineering with synthetic fuels and agricultural and forestry products. Nondegree departments, too, were engaged in keeping their curricula up to date. Clearly, the immediate postwar period was a

The typical Miner—and there were many in that role.

twofold opportunity to review and improve what Mines had been doing and to move into new areas of teaching and research.

Maintaining summer school at nine weeks and allotting six weeks for summer field courses enabled the school to operate throughout the year, thus reducing some of the pressure on classroom space. The 1947 summer school boasted students from every state but Rhode Island and from fourteen foreign countries, proof again of Mines' strong national and international reputation. Fortunately, Mines was able to purchase at considerable discount much furniture and equipment from the Office of Price Administration to ready the temporary buildings for the 1947 fall term. The Experimental Plant, which had been active in efforts relating to the war industry, resumed its normal testing and research for the minerals industries. Field courses were resumed at the experimental mine in Idaho Springs, at Wild Horse Park near Pueblo, and at the petroleum production camp at Rangely.

The Mines ROTC advanced course (last two years), discontinued during the war, was reactivated, with the welcome addition of Col. Wendell W. Fertig, famed Philippine guerrilla leader, as professor of military science. From its inception in 1919, Mines' ROTC program was a vital part of the curriculum and gave the school an exceptionally strong reputation in preparing officers for the Army Corps of Engineers. Since the basic course of the program was required, it grew with the school, as did the advanced course. Because the Mines ROTC unit graduated the largest number of reserve officers for the Corps of Engineers of all other units, Maj. Gen. Eugene Reybold, at the beginning of World War II, referred to it as the "backbone of the Corps of Engineers."

Students in the reactivated advanced course received a uniform allowance of $84 and subsistence of $27 per month. Also, Distinguished Military graduates became eligible for Regular Army commissions in the Corps of Engineers. In 1949, the Department of the Army initiated significant changes in the ROTC

Single students, too, faced a housing shortage.

curriculum and added new subjects to both the basic and advanced courses. Basic students would study Hand Tools and Rigging, Camouflage, Explosives and Demolitions, Mines and Booby Traps, and History and Traditions of the Corps of Engineers. Advanced students would study Military Roads and Runways, Organization of Combat Divisions, Vehicle Operation and Maintenance, Engineer Support for the Air Force, Engineer Support for the Communication Zone, Engineer Support for the Field Type Army, and Motor Movements. Modern war demanded a Corps of Engineers equipped and trained to handle situations that previously were only imagined.

The ROTC program nationally had without question proved itself in World War II one of the most valuable training components in the United States military. It would do so again in the Korean War. Mines, of course, was among the leaders in terms of numbers of ROTC graduates who fought in both wars. Norm Zehr '52, assistant editor of, and contributor to, *The Encyclopedia of the Korean War*, and director of the Mines Alumni Association during the 1980s and 1990s, listed 391 Miners who participated in the Korean War. Zehr himself was an army aviator in that conflict.

II

In April 1950, a notable administrative switch occurred at Mines. Parker resigned his presidency to return to industry and to become a member of the Mines Board of Trustees, for which he would serve as president for twelve of his nineteen years on the board; and Dr. John W Vanderwilt left his position as a trustee to succeed Parker as president of the school. A native of Iowa and a well-known geologist, Vanderwilt held a B.A. from William Penn College in Iowa, an M.A. from the University of Michigan, and a Ph.D. from Harvard. He served on the Board of Experts of the U.S. Bureau of Reclamation and as a consultant for various mining companies. He taught at both the University of Colorado and at Mines for short periods. Here were two men, each serving as professor, president, and trustee at Mines, literally exchanging the positions of president and trustee.

In leaving his post of president, Parker took the opportunity to comment more frankly than he could while in office about the school's problems and the responsibility of the state in helping to solve them. In "Mines Deserves Better State Backing," he decried the lack of awareness of Coloradoans of the many benefits and contributions they enjoyed because of the location of Mines and its support of the minerals industries:

> With these benefits in mind, it is most disheartening to look back on the history of the Colorado School of Mines and realize that for more than half a century the contribution of the state to the support of the school has been most meager. It is only in the last few years that a beginning has been made toward providing the support that it should have. The success of the school in the face of these

Wives of students found companionship in the Dames Club.

Chapter 6: A Decade of Change

The Power House kept Mines functioning.

financial limitations has resulted largely from the loyalty of its faculty, which always has included authorities in their fields who could have substantially increased their earnings in private industry. (MM February 1950)

Parker stressed the legislature's need to realize that the cost of an engineering education rivals that of a medical education in terms of laboratories and equipment, as well as the desirability of small classes. Pointing out that the industrial concerns of Colorado have largely left the support of the school to out-of-state corporations, he notes the state's industries would do well "to consider the advisability of providing support for Colorado's world-famous educational institution which supplies much of the trained personnel required for the mineral industry."

Vanderwilt faced many of the same challenges as did Parker regarding state funding. Mines was still in a growth pattern, with classrooms, laboratories, and faculty utilized to the maximum. Enrollment for the 1949–50 academic year was 1,134, a slight decrease from the previous year but still a 50 percent increase over prewar attendance. Degrees offered were Geological Engineer, Geophysical Engineer (Geophysics was now a separate department), Metallurgical Engineer, Engineer of Mines,

Petroleum Engineer, and Petroleum Refining Engineer. The housing shortage remained a serious problem, as faculty, married students, and single students scrambled for accommodations. Veterans and their families were still making do with trailers and converted army barracks for housing, and faculty members were often living considerable distances from the campus. While construction of dormitories was still on the back burner, work started on Coolbaugh Hall to house Chemistry, as did planning for a new Petroleum building.

World War II and the ensuing cold war had considerable impact on the American mining industry. In an article titled "Can Mining Take It?" Robert S. Palmer, executive manager and secretary of the Colorado Mining Association, noted that although the American system "outproduced and whipped" our enemies, "the battle of the home front continues. Disturbing elements are at work; our nation faces a future veiled and uncertain." Palmer cited various mining regulations set by the government that tended to weaken industries both during and after the war. To the charge that the

---

Golden Daze

...there I was...stirring six-cent coffee

...with my little finger...chalking a cue stick

...catching the sun...oiling my slide rule

...quizzes ahead...quizzes behind...ah,

this Colorado School of Quizzes...ah, life

...thank God it's Friday...the easy life

...that's for me...dances...parties...

stags...drags...what all...and as the

sun sinks slowly in the west, some people

graduate...

Anonymous

> ## Traditions (from Jack Haley '48)
>
> The class of '48 (most of whom started in 1941) experienced many of the school's traditions and contributed to the demise of some.
>
> Hazing started when we set foot on campus and continued through the all-school gauntlet just before Christmas vacation. Fraternity pledges received additional hazing that culminated in Hell Week at the end of the first semester.
>
> We wore green miner's caps and obeyed upperclassmen's instructions to "button up" (hands on top of head) and "sound off" (beat weekend athletic foe). Certain areas of the campus were taboo, and we had to respond to questions on school history and traditions. Violators were summoned before a kangaroo court.
>
> Sophomores ran us through two gauntlets a week, one when the class emerged from Doc Ward's chemistry lecture and another when we left the Armory after stacking our rifles following ROTC drill. There was only one exit from each of these. Buttocks were to be swatted with belts; however, some sadists used the buckle end.
>
> Fraternity initiation required keeping a paddle available, ingress and egress through basement windows, and treasure hunts. The treasures included spittoons from Union Station, souvenirs from DU fraternity

Colorado Mining Association has had political tendencies, he responded, "If our legislative bodies are to determine our future, then we of necessity must work with legislative bodies. If executive departments are to decide our right to mine, then we as a combined group of mining people must work with the executive branch of our government." Palmer represented the common business view that the less government intervention the better. He even suggested, no doubt facetiously, that Colorado mining interests "would get more concessions from the Federal Government if it would secede from the Union and be given the right to send diplomatic representatives to Washington, who could set up an Embassy and deal across the table with Uncle Sam" (MM February 1946). Palmer surely must have realized that he was fighting a rearguard action. The decline of the mineral industries in the United States had begun years before, a decline that held significant curricular implications for Mines.

Vanderwilt, in his 1948–50 Biennial Report, pointed to some of the curricular revisions underway to address changes in the mineral industries:

> Changes in the curriculum were made where necessary in order that the courses would meet more exactly the needs of young engineers entering the mineral industries in administration, operation and technology. This is especially necessary here, for in no other field have so many changes and developments occurred in recent years. This biennium proved that changes providing for specialized study in coal mining and for separate departments of petroleum engineering and petroleum refining were well warranted.

Recognizing the nationwide shortage of mineral engineers caused by World War II and then by the Korean War, Vanderwilt urged the state to provide houses, and mementos from the red-light district. Pledges had to memorize pertinent facts, founders' names, and some pure nonsense.

This class put one of the silver stakes through the heart of Senior Day. Starting with kangaroo court, we summoned the professor for reinforced concrete and set his feet in a tub of cement. After the traditional seniors-faculty softball game with a keg at every base, someone came up with a flatbed to convey beer and passengers, and we headed for Colorado Women's College in East Denver with several following in cars. When two of Denver's finest pulled alongside to ask just what the hell was going on, someone in the flatbed's cab elected to add to their comfort by heaving a pitcher of beer at one through the open window. The recipient bailed out of the cruiser, and the *News* photographed class president Robin Corbett explaining everything while mopping the officer's uniform with a handkerchief from the breast pocket of his pj's. This wasn't enough to satisfy some dimwits who later planted a few sticks of dynamite up a couple of trees on the campus of Colorado College in Colorado Springs.

Hazing served to bring freshmen together and to organize them for survival while indoctrinating them with school spirit, history, and traditions. Like many worthwhile activities, certain parts got out of hand. Their time had come.

the necessary funds to keep the school operating at its capacity: "Industry and the Armed Forces both look to Mines as to no other college to supply the men they need to find the oil, metals, and strategic mineral deposits and to build vital bridges and roads for an advancing army."

III

In the early 1950s, America's engineering schools were not producing enough graduates to keep up with the nation's increased demands. In "A Helluva Shortage of Engineers," Lawrence Lessing described the scramble of industries and government competing for graduates. Virtually every category of engineering was in the market. Lessing recognized "flesh-peddling" agencies and dummy engineering companies contacting batches of engineers and then hiring them out to hungry industries at inflated salaries and pocketing the difference. Monthly salaries, according to Lessing, ranged from $290 to $325 for bachelor's degrees, $350 to $400 for master's degrees, and $450 to $750 for doctoral degrees. "The engineering student," said Lessing, "suddenly found himself rivaling the football star as the glamour boy of the campus." The problem, moreover, was only going to get worse.

Lessing also addressed the perennial question of whether engineering is a profession. The engineer, he complains, "is still confused with the railroad engine men and other begrimed machine tenders by many people" (*Fortune Magazine* September 1951). Since 1951, formal associations have been established for accreditation and certification; laws to guard the standards and practices of engineering have been established by all the states.

C. M. Hathaway, in "Engineering Must Have Prestige," blamed the apparent lack of prestige granted engineering as a profession stemming from

These Tau Beta Pi men hardly resemble "ghosts."

the public's lack of understanding of what engineers really do. He drew an apt distinction between medical doctors and lawyers on the one hand and engineers on the other. Unlike engineers, doctors and lawyers receive publicity for working in the public interest and thus receive publicity necessary for prestige: "Instead of working on his client's health or prestige, the engineer works on machines or devices or constructions that will affect the public health or prestige." If the engineer is to receive the public prominence he deserves, according to Hathaway, he must "step forward to take the place of leadership in a technical world that is rightfully his, a place that only the engineer can fill as it should be filled" (*Engineer's Bulletin* June 1951).

The foundation of any profession is the education required of its members. The American Society of Engineering Education (ASEE) designated 1953–54 as a year of evaluation, ranging from public school preparation through the undergraduate and graduate programs. In his presidential address at the ASEE convention in 1954, L. E. Grinter discussed these areas of evaluation, beginning with a castigation of the progressive education movement that had displaced rigorous study of the three Rs in favor of preparation for life through social study and group activity. The greatest concern to engineering teachers, he emphasized, "is not the loss of subject matter content that the high school student presents for college entrance, but the lack of development of the rigorous thought processes that accompany the study of mathematical analysis, English expression, and the logical method of science."

Grinter also criticized engineering educators who attempted to produce well-rounded graduates by trying to accomplish everything for every student, a goal that he said has resulted in a "fixed curriculum for every student after an initial choice has been exercised." The problem was time. How all the material could be squeezed into a four-year course was a question that to this day has not been answered definitively. Two possibilities have been considered and reconsidered: to increase the undergraduate course to five years or to require the master's degree for engineering certification. Grinter argued that "the profession of engineering requires the same elements of education as all other professions, but with greater emphasis on creativity" in order to "weld art of practice with basic science in an approach so fundamental that it will level the wall of narrow professional specialization which has been raised by the pride of craftsmanship" (MM July 1954).

In the early 1950s, the field of engineering was examined from a number of angles: the shortage of

The Fifty-fifth Associate Engineers Company was made up of many Mines men who served in the Korean War.

Chapter 6: A Decade of Change

The 1967 Mines Board of Trustees. Left to right: Ted Stockmar, Ben Parker, Russell Volk, Edward Eisenach, and William Alexander.

engineers, the effort to define engineering in professional terms, and the ever-present question of the education of engineers—all of which had an impact on schools of engineering as the century moved into its second half. However defined or described, engineering was a profession, and the demands put upon it to build and produce at an ever-increasing pace indicated that its status would continue to rise on the scales of respect and necessity. If Mines had its spirit, so too did the profession of engineering in general, whether it was seeking the resources buried in the depths of the earth or soaring into the sky toward new worlds to explore. That spirit was, and is, not simply the product of mathematical formulae, football games, or campus traditions but the product of individuals with a vision both practical and ideal. When Hart Crane celebrates the Brooklyn Bridge both physically and metaphysically in his poem "The Bridge," he sees the masses of people moving daily between Brooklyn and Manhattan as a material testament to the skills of the designers and builders of that bridge. When the same bridge awakens a sense of wonder and awe in the many who gaze at it from a distance, it stretches beyond the practical and becomes a symbol of the human imagination and creativity of those designers and builders, proving after all that engineering and poetry can be one.

IV

Vanderwilt's annual progress reports during the 1950s reflect Mines' efforts to remain in the mainstream of engineering education. The 1951–52 report covers briefly every department of the school in terms of both enrollment and degrees granted. Geology continued as Mines' largest department, enrolling 191 students in the fall and 180 in the spring. Fifty-seven candidates received Geological Engineering degrees; six, Master of Science degrees; and four, Doctor of Science degrees. The faculty conducted 2,730 class periods and 2,091 laboratory sessions in the two options of mining-geology and petroleum-geology. Geophysics, having received degree status in 1949, continued its impressive growth, with over 150 students enrolled in 972 classroom sessions and 428 laboratory sessions. Of the thirty-four geophysics courses offered, eighteen were at the graduate level.

Metallurgy and Mining were also advancing. Study in the former was divided into four options: mineral dressing, production metallurgy, physical metallurgy, and metallurgical research. The department

John W Vanderwilt moved from trustee to president, 1950.

129

No parabolic skis for these gentlemen.

conducted 1,140 classroom sessions and 221 laboratory sessions, serving 276 undergraduates and 45 graduate students during the fall semester and 475 undergraduates and 17 graduate students during the spring semester. Mining enrolled 126 students in the metal-mining option and 22 in the coal-mining option during the fall term. The figures for the spring term were 121 and 17, respectively. Mining students used the newest types of mining equipment at the experimental mine at Idaho Springs. With increasing demand for petroleum engineers in the areas of oil and gas, the Department of Petroleum Engineering had 110 students enrolled in the fall, and 105 in the spring, in 272 classroom sessions and 162 laboratory sessions. The curriculum was divided into four sections: drilling, production, fluid flow, and business of the oil industry. The Department of Petroleum Refining, already considered one of the best of its kind, enrolled 84 students for the year and conducted 926 classroom and 187 laboratory sessions. Nondegree departments that contributed significantly to the curriculum were Chemistry, Civil Engineering, Descriptive Geometry, Economics, Electrical Engineering, English, Mathematics, Mechanical Engineering, Military Science, Physics, and Physical Training and Athletics.

Enrollment at the school for 1951–52, affected somewhat by the Korean War, was 852 in the fall and 835 in the spring. Of these, 96 were graduate students in the fall and 99 in the spring. Another factor indicative that Mines was returning to a prewar student body level was the declining veteran enrollment. While still fairly heavy at the upper levels, veteran enrollment was falling off considerably at the freshman level. Veteran enrollment was 215 in the fall (8 freshmen) and 182 in the spring (10 freshmen). Mines in 1951–52 had no trouble placing its graduates and was looking eagerly toward expanding in numbers of students and in new physical facilities.

Building a reputation by dropping students for academic reasons was a charge that Mines had often faced. H. D. Burdick, director of Admissions, discussed this question in an article in MM. Citing a study by the Educational Testing Service, Burdick found "some interesting and almost startling facts." The study shows that in 1952 of the nonveteran students who enrolled in engineering colleges across the nation, 33 percent completed their degree requirements in four years. At Mines, however, 42 percent completed their degree requirements in four years. Mines, moreover, ranked in the 70th percentile in number of students graduated in four years. Burdick concluded:

> The Colorado School of Mines does *not* fail as many as the average engineering school of the

Following the leader among the footprints of a winter day.

> ## Mojave—or Opened by Mistake
> ### (from Jack Haley '48)
>
> William T. (Mojave) Griswold, my first roommate, was a mild-mannered, soft-spoken, serious individual who had spent a lot of time on the high desert of southern California. He owned a black 1927 Lincoln Town Car (one of the few vehicles on campus) with speaker tubes and a sliding glass behind the driver. A hobby was loading ammunition and shooting his twin Colt 45s. His marksmanship stood him in good stead later when he flew P-40s and P-51s over the Hump and in the Pacific theater. When "The Ghost Walked" (Theta Beta Tau pledging ritual) that year, they also had a list of those who they believed failed to demonstrate proper respect, and Mojave was on it. Their specialty was treating tender parts of the body with a mixture of catsup and carbon tetrachloride. Our door was bolted shut, but they kicked it in. Rushing inside, they found themselves looking down the barrels of those 45s. The Ghost vanished and Mojave was off the list—permanently.

nation, and this is probably due to the selection used in admitting high school graduates and others to the School. In 1951, the entrance requirements to Mines were tightened to exclude those who did not have the preparation necessary for successful scholastic work. Before 1951, any applicant ranking below the required upper third of his high school class needed only two letters of recommendation to be admitted. Since 1951, the applicant, not in the upper third of his class, must score above average on the aptitude test of the College Entrance Examination Board. (August 1953)

As part of Mines' 1951 curriculum evaluation, its Curriculum Committee undertook an extensive study of the curriculum, particularly of the first two years. To ensure that all students be well prepared in basic areas of study, the committee recommended, and the faculty accepted, a uniform course of study for the freshman and sophomore years that included general chemistry, engineering drawing, descriptive geometry, English composition, elements of geology, historical geology, freshman mathematics, quantitative analysis, organic chemistry, mineralogy, crystallography, calculus, general physics, military science and tactics, and physical training. Also, the maximum credit-hour load per semester was reduced to twenty-one. Courses for the freshman and sophomore years totaled nineteen or twenty hours per semester. Summer field sessions added another seven or eight hours.

While Mines had from time to time examined and assessed its educational efforts, it had not yet embarked on serious long-range planning. With the end of the school's first century approaching, the Board of Trustees, the administration, the faculty, and the alumni joined forces to create a long-range plan stressing the importance of increasing state and private support to strengthen the school's overall educational program in both the short and long term. A booklet titled "The Hundred Year Horizon," published in 1956, outlined the plan. At a dinner at the University Club in Denver on 17 February of that year with Governor Dan Thornton as an honored guest, Vanderwilt presented the plan. Financial goals aimed for $445,000 annually to augment state funds and $20 million in long-term

The Edgar Mine

The Edgar Mine, located near Idaho Springs, Colorado, once a producing minerals mine (gold, silver, lead, and copper), was acquired by Mines in 1921 through a long-term lease from the North American Mining Company and, eventually, was given the formal title of Colorado School of Mines Experimental Mine. Two tunnels, the Miami and the Big Five, and one shaft, the Edgar, provide entrance to the mine. Made up of over four thousand feet of tunnels, the mine, with its safe rock formations and appropriate underground workings, provides mining students research and practical training in underground surveying, blasting, ventilation, unit operations, and safety. Facilities at the area consist of a main building with an office, a drafting room, a computing room, a storage room, and bathrooms; a blacksmith shop; and a compressor and machinery house. Hundreds of thousands of dollars worth of equipment, much of it contributed by manufacturers of mining machinery, have made what could have become a ghost mine an invaluable educational facility for mining students.

In the April 1929 *Compressed Air Magazine*, C. H. Vivian, in his descriptions of various experimental mines across the United States, described the Edgar Mine in glowing terms:

> Thorough practical training is given in all phases of mine surveying. In addition to the underground facilities, the surface topography is favorable to instructions in mineral-land surveying. Lode claims are laid out with discovery shafts on actual veins and tied in with government section corners, mineral monuments, and triangulation stations. Adjacent to the mine is Clear Creek—the flanking ground being admirably suited for practical work in placer and mill-site surveying. Conditions in the workings are satisfactory for mine sampling and valuation work in mining geology, as well as for all actual mining operations.

In addition to serving as a necessary laboratory for prospective mining engineers, the mine carries out cooperative research for numerous academic, government, and industry groups, including the Bureau of Mines and the U.S. Army. In addition to its significant role in mining education and research, the Edgar Mine, a tourist attraction, offers tours from late May through late August.

Pictured above: Entrance to the school mine.

capital funds from both state and private sources. The Horizon Plan operated under the auspices of the Board of Trustees and a revitalized Colorado School of Mines Foundation. Mines had now entered the financial derby of American colleges and universities, a race in which presidents become fundraisers as well as educators, with increasing emphasis on the first.

The educational aim of the plan proposed a number of incentives that would help Mines attract and keep a strong faculty in an increasingly competitive marketplace. Salaries and benefits, including a sabbatical-leave program, would be increased, and teaching loads reduced. Improved physical facilities and more clerical and technical staff would encourage and support research. These steps, tentative as they might seem now, indicated the right direction and set Mines into a competitive mode.

The Colorado School of Mines Research Foundation helped to meet the needs of the minerals industries. Housed in the Experimental Plant, the foundation was established in 1948 as a nonprofit corporation to carry out research in various fields of the mineral industries for individuals, industrial organizations, or government agencies. In the few years of its existence, this foundation became a leader in minerals research, exploiting some of the most extensive facilities for research in geology, geophysics, metallurgy, mining, petroleum production, and petroleum refining. During the 1951–52 academic year, fifteen full-time and fifty-eight part-time employees among Mines faculty completed one hundred research projects. Income from contracts negotiated for sponsored research covered the foundation's operating costs.

The academic buildings Mines so sorely needed finally began to appear in the 1950s. Coolbaugh Hall (Chemistry), completed in 1952 at a cost of $1.05 million, was dedicated on 4 May 1953. The first completely new classroom and laboratory

Oh, it's dark as a dungeon way down in the mine.

The entrance to the Mines museum at the west wing of Berthoud Hall.

building constructed since 1940, Coolbaugh Hall had 65,000 square feet of space on three floors. It boasted thirty classrooms, eight laboratories, fourteen faculty offices, and an auditorium seating 306. Old Chemistry, constructed in 1879 for $7,500, was enlarged in 1882 and again in 1890. So much for inflation. Lester C. Thomas, president of the Board of Trustees, noted at the dedication that Coolbaugh Hall "was constructed as a lasting memorial to a great man, Melville Fuller Coolbaugh, whose intimate association with the history and ideals of the Colorado School of Mines made it possible."

Alderson Hall, completed in 1953, was named for former president Victor C. Alderson, who established the first petroleum-engineering curriculum at Mines. Having cost and square footage comparable to Coolbaugh Hall, Alderson Hall housed the departments of Petroleum and Petroleum Refining. Among all the necessary laboratories and classrooms was a complete simulation of an oil rig of one hundred feet with platforms at each floor level, thought to be the only indoor oil well of the time. The English Department occupied the north wing of the first floor.

Housing, while not so tight as immediately following the war, was still a problem for both faculty and students. Mines Park, constructed in 1939 for faculty housing, consisted of eight efficiency apartments, eight one-bedroom apartments, eight two-bedroom apartments, and six two-bedroom houses. The vacancy rate was extremely low, and lengthy waiting lists were common. These units were popular with faculty members until the mid-1990s when they at last were torn down to make way for modern student housing. On Lookout Mountain Highway, Prospector Park, the principal student housing project owned and operated by the school, provided seventy-six one- and two-bedroom apartments for married veterans. These units, too, were in high demand and kept long waiting lists, largely because the rent was only $35 a month. Located next to Brooks Field, trailers were still available for married students, veterans or not. More housing, however, was needed.

The first dormitory on the campus, Bradford Hall, completed in 1954 at a cost of $250,000, was named for A. Hartwell Bradford '09. Bradford was a pioneer in the development of natural gas extraction and founded the West Coast Refining Company and the Arrowhead Oil Company, as well as Bradford Brothers, a family-owned investment company. The family was also active in developing the California citrus industry.

V

Mines was not founded as a male-only institution. Indeed, the Territorial Act of 1874 stipulated clearly that the school was to be "open to any inhabitant of the Territory of Colorado without regard to sex or color." A number of women enrolled over the years, but only three actually received degrees prior to 1949, the year Jacquelyn Borthick was graduated with the degree of Petroleum Refining Engineer. *A Century of Women at Mines*, an excellent history of women at Mines published in July 1999, describes Jacquelyn Borthick as well liked by her fellow classmates and often participating in extracurricular activities. She was chosen Homecoming Queen by popular vote in 1948. According to *A Century of Women at Mines*, she left her home in Cherry Hills to room in Golden, where she solved the problem of meals by arranging to share meals with the cook at the Beta Theta Pi house. She was not allowed in the dining room with the men. Upon graduation, she married a classmate, Rex E. Kircher '49, also a Petroleum Refining Engineer. Nancy Easley followed Borthick in 1952. Though she attended Mines for only two years before transferring to the University of Colorado, she brought national acclaim to the school via a full-page picture and cover story in the 24 November 1952 issue of *Life* magazine. The striking photograph, titled "All Alone with 900 Boys," showed Nancy at the head

Facing the door of Stratton Hall.

of some 750 of those boys as they flowed down the steps of Guggenheim.

While the ratio of males to females remained overwhelmingly unbalanced for some years, the social dynamics of Mines soon began a gradual but dramatic change. In the early 1950s, however, Mines was Mines and still essentially male; its male-oriented traditions were not to bow out gracefully. Thus, when the headline of the *Oredigger* of 20 May 1952 read SENIOR DAY IS ABOLISHED, there was understandably some consternation among the students. Senior Day had indeed been abolished in the past, but with no lasting success. One of the more cherished and troublesome traditions on the Mines campus, Senior Day was the springtime bash that made more than one president quiver with apprehension. Discussion between the administration and the seniors prior to the appointed day did not result in any improvement, at least in the eyes of the first party. Thus, a letter dated 9 May 1952 was sent to all Mines students and their parents, as well as to alumni, from the president of the Board of Trustees and Vanderwilt declaring the abolition of Senior Day and the reasons therefor, with the stipulation that students failing to comply with the regulation would be expelled and would not be graduated. An echo of the past, to be sure, but this time it was for keeps. Some seniors attempted to rescind the ruling the following fall, but Vanderwilt's response was that the ruling was necessary for the good of the school. Although there were efforts to revive it in later

A Chevy Named Buttercup
(from Hugh W. Evans '49)

I remember Buttercup. She was a grey, old, 1937 Chevy sedan when I bought her from Tony DeMoraes for, I think, $150. I painted her yellow and gave her her beautiful name. One night coming off Lookout Mountain, it was so [cold] we knocked her roof out and built a small fire from the kindling gained. From then on the sun and moon had access to her innards. It was lovely.

With the roof out, Buttercup became a scout car for the Lookout Mountain School for Boys where I worked as an orderly in the infirmary at night. Every so often, a boy would run away and we were off after him with an assistant standing in the back seat looking for the escapee. More than once, we met with success in the chase.

Buttercup met her demise in Salt Lake City about a block from Bob Reeder's home. I had offered Bob a free ride home. He has never forgiven me. I was headed for California but had to take the bus the last 752 miles. It only took us twenty-four hours from Mines to Salt Lake City. We were picked up in Longmont for no taillight, so I bought a red flashlight to hang on the tail end. The lights went out in Wyoming, and I could not go fast enough to keep up with the truck ahead, so we stopped for the night by the road. The mosquitoes had a heyday. At first light, Buttercup would not start, so we hailed a P.I.E. truck for a pull off. Off came the bumper! A second try worked, and we were off and running—sort of. Well, to make a long story short, after two flat tires and a warped axle, the drive shaft gave out just as Reeder's home came in sight. The scrap dealer gave me enough to buy my bus ticket home. Check with Reeder for the details.

Time for a smiling break in concentration in Arthur Lakes Library.

years, Senior Day, as it once was, lives now only in the memories of alumni who knew it and in the scattered copies of its special newspapers, themselves now collectors' items.

The Gauntlet, another time-honored tradition in which freshmen were forced to run between two rows of belt-swinging upperclassmen, was soon to be made a matter of choice on the part of freshmen, even though an all-school survey taken in the spring of 1952 (see the *Oredigger* of 15 April 1952) showed 80 percent of the student body favored keeping this tradition. One suggestion was made that Blue Key should keep a closer watch on types of belts used and how they were brought to bear on the freshmen. Women, if they participated, were escorted, and no doubt shielded, by Blue Key through the Gauntlet.

The many veterans who populated American colleges and universities during this period were often ambivalent regarding long-standing traditions that originated in the minds of younger students of years gone by. Some wanted to participate fully in undergraduate life, but many who were married, perhaps with children, had little time or interest in traditions they saw as juvenile. The mere factor of growth in enrollment also played a role. As colleges and universities grew, traditions tended to become less important and even to disappear. The Mines spirit, to be sure, would remain strong, but it would be defined and exemplified differently.

CHAPTER 7

Still Achieving, Still Pursuing

The challenges of life and the needs of society defy traditional boundaries, and the collective expertise of the campus is an asset of growing value that is widely acknowledged, though not yet fully utilized.

—Frank H. T. Rhodes

Despite periodic anxiety from the cold war, America in the later 1950s emerged as a military and economic superpower. Domestically, technological advances and mass marketing drove the economic engine toward the twenty-first century. The specter of overproduction, however, constantly lurked in the background and regularly provided "booms" followed by "busts." What might be seen as an exception was the birthrate. Indeed, the 1948 Berlin Airlift was outclassed by the 1950s flights of the stork in America. Although the famous American Western frontier, according to Frederick Jackson Turner, may have closed in 1890, America has never been without some kind of frontier, be it the New Deal, the New Frontier, the Great Society, or Space; and one of the more influential frontiers of the 1950s was that of suburbia. Young married couples, whose parents, more often than not, could only dream of owning their own home, now purchased picture-window houses in the $20,000 range throughout suburbs that blossomed profusely around cities, the centers of which soon became economically and socially distressed areas. The streets of this new frontier were regularly patrolled by baby strollers with occupants who, reared on Dr. Spock, the Mouseketeers, and the Golden Arches, soon changed virtually all of the nation's social dynamics, not the least of which was education from kindergarten to the university.

I

Mines' enrollment for the 1956–57 fall term was relatively strong at 1,162, representing forty-six states and thirty-seven foreign countries, the latter making up 10 percent of the student population. At the May commencement, 161 undergraduate and 17 graduate degrees were awarded. Metallurgy was the most popular option, followed by Petroleum Engineering and Geophysics. Adding to the general feeling of optimism on the campus, the Engineering Council for Professional Development (ECPD), following its visit,

Opposite page: Pep rallies, bonfires, and romance—the perfect collegiate scene.

Alderson Hall serves well as a tribute to a controversial president who served Mines well.

accredited all six of the degree-granting departments and indicated that all departments of the school met the standards set for engineering colleges. According to Vanderwilt's 1956–57 Annual Progress Report, the ECPD was particularly impressed with the thorough training given Mines students and the "professionally competent, creative-minded, and enthusiastic faculty."

As glowing as Vanderwilt's report was, another accreditation crisis loomed. In 1929, Mines had gained membership and accreditation in the North Central Association of Colleges and Secondary Schools, one of the nation's influential regional accrediting associations. Lack of such accreditation could cost an institution recognition of its diplomas and cause difficulty in the transfer of its credits. In 1937, however, the Mines catalog curiously made no mention of such membership. Whether the school felt that its specialized mission did not correspond with North Central's accreditation requirements and simply let its membership lapse or whether North Central dropped Mines for some other reason remains unclear. Dean of Faculty Truman Kuhn realized the importance of North Central recognition and wrote to the secretary of the association on 19 November 1957 requesting re-establishment of membership and accreditation, noting that the lapse must have been a mistake. The procedure, however, was not simple. Mines was required first to present a self-study that covered its entire organization and operation, the appropriateness of its curriculum, and the quality of its faculty and students. If the self-study proved promising, North Central would send a visiting committee for a more detailed examination that would conclude with a recommendation to accept or refuse Mines' request. A steering committee of seven administrators and faculty members led the study, utilizing materials from previous studies, committee reports, faculty conferences, and special departmental meetings called to implement the work of the steering committee. The result was a detailed 200-page presentation titled "Colorado School of Mines Self-Survey Report," printed and bound by the Mines printing department and sent with high regard to North Central.

While recognizing that Mines' educational mission of preparing quality engineers for the mineral industries was clearly defined and being carried out expeditiously, the committee evaluating the report concerned itself with low faculty salaries, faculty competence, and weakness in general education requirements. Because engineering schools had to compete with industry and each other for quality faculty members, the committee emphasized that, to secure and retain such people, Mines would have to increase salaries substantially. Further, "an examination of the faculty list in the general catalog indicates that a number of the faculty are deficient in graduate preparation in their respective fields to maintain high standards in research and graduate

It appears that there was a shortage of rifles for this parade.

Like Mines, Coors of the 1950s had come a long way as a vital part of the Golden (and Colorado) scene.

instruction." The committee agreed unanimously, moreover, that Mines needed to strengthen its general education program, a view one of its members summarized:

> Unless I am much mistaken, the engineer of the future ought to have certain things expected of him intellectually which the Colorado School of Mines is not working with at all at the moment. By this I do not mean the usual stucco of general education, but some means of jarring young men of considerable technical competence into the realization that technical power is the servant of a society, not its center. Certainly I would not be willing to approve as good education the graduation of groups of technicians with no idea of what their own center or the center of their society really was.

This comment echoes a perennial question faced by many schools of engineering: What is the correct balance between courses in technology and science and those in liberal arts? Despite the reservations, the committee unanimously recommended that North Central consider Mines' application for reinstatement.

Two examiners visited Mines on 10 and 11 February 1959, G. A. Hawkins, a former Mines student, though not an alumnus, and J. W. Nelson. Following a thorough study of virtually every aspect of the school, the examiners recognized a number of strengths:

> The Colorado School of Mines is a strong and distinctive educational institution. Through the clear-cut limitation of its objective, it has been able to attain renown as a center of mineral engineering education. It has maintained, and still maintains, high academic standards. Its graduates are known abroad, and men from many parts of the world come to its campus for their training. The present staff is efficient in teaching and in research activities. The support which the state has given the school, especially in recent years, is

Up, up, and away!

satisfactory, although that support should be increased. The School of Mines' position among the institutions of higher education in Colorado is a major one.

In short, Mines was doing a first-rate job in its educational mission. Hawkins and Nelson, however, expressed the same strong caveat as the previous committee:

> The one major weakness of the School is its inadequate program of general education. The examiners do not see how this weakness can be eliminated without changing to a five-year curriculum. But whatever the solution, they must point to the lack of sufficient liberal studies as a defect which, from the standpoint of the North Central Association, cannot be overlooked and which cannot be compensated for by the obvious elements of strength which the institution as a whole displays.

The examiners recommended that Mines not be accredited immediately but that "it should be invited to apply for membership again in two or three years if it is convinced that the weakness in general education has been eliminated."

With this both welcome and disappointing news, Vanderwilt began a campaign to gain North Central's blessing as soon as possible. He went to Chicago to meet with the full Accreditation Committee of North Central on 20 April 1959. In a letter of 18 September 1959 to Norman Burns, the association's secretary, he stressed that during the Chicago meeting he had not been given a specific idea of what kind of general education requirement North Central demanded:

> It was my impression that committee members were not in full agreement on either point [subject matter or amount of general education], and statements as to minimum requirements were in such general terms as to be not at all clear to me. It was finally suggested that I consider withdrawing the application of the school and in this manner avoid the embarrassment should accreditation be denied because of deficiencies in general education.

Later in the same letter, Vanderwilt deftly placed at least some blame on North Central for vagueness in its own informational material:

> Frankly, I must confess that even with the advantage of hindsight, we have been unable to find the answers in the literature available to our staff for its use in connection with our self-study report. The information that still is lacking and which we would like very much to have as soon as convenient is (1) How is general education defined by NCA in connection with accreditation of engineering curricula such as we have at the Colorado School of Mines? and (2) How are the minimum standards for general education in engineering curricula defined?

Burns, in his reply of 21 September 1959, noted:

> As I am sure you will understand, a concept such as this cannot be expressed in terms of minimum number of hours of credit in several fields of subject matter. The North Central Association expects an institution to offer a well-organized program of liberal studies which includes work in the major areas of human knowledge. We cannot, nor would we attempt to, prescribe the program which an institution should adopt. That is the task of the institution.

The above exchange, not to mention the entire process and purpose of accreditation, illustrates the differences between the perspectives of engineering and liberal arts. Not a matter of right or wrong or what can be quantified and what cannot, the question is how to define what is meant by an "educated person."

On 24 November 1959, the Mines faculty adopted a multifaceted curriculum revision endorsed by the administration and the Board of Trustees that supported additions to humanities and social sciences. The revisions were as follows:

1. Required courses in the humanities and social sciences increased from twenty-one to twenty-seven hours, including three hours in rhetoric and orientation and six hours of economics.

2. Mathematics in the lower division reduced from twenty to seventeen hours, with calculus introduced in the freshman year.

3. Physics in the lower division increased from ten to thirteen hours.

4. Chemistry in the lower division reduced from twenty-six to sixteen hours, with the reductions principally in analytical chemistry and organic chemistry. Physical chemistry moved from the junior year to the sophomore year.

It may not be New Year's, but the Orediggers play in a setting more impressive than many of the notable "bowls."

Transition: Colorado School of Mines
(excerpts from a column by Lee Turner '70 in the *Technocrat* 8 November 1969)

To alumni returning for Homecoming '69, and perhaps even to seniors who came here as freshmen, the Colorado School of Mines is a changed place. Homecoming '69 seems an appropriate time to look at the old and the new, to analyze some of the changes that have occurred in the past few years, and to place them in perspective.

I came to the school in September of 1965, just a few months after the first ground troops were sent to Vietnam by President Johnson. But Mines freshmen were not concerned about the war. We were busy learning our Freshman Bible, wearing our clothes backward, and thinking about the "Ghost of Theta Tau." In '65, we heard about the good old days from the upperclassmen, about the rough-and-tumble mining days when Mines' men were men and their women were proud of it. But in '65, times were changing, as time has a habit of doing. We heard about the demise of "running the gauntlet." We heard about how easy we had it, having agitation for only 2 weeks.

In the fall of '65, the school's enrollment reached an all-time high and agitation was cut in half. The administration proposed a degree change and the bells of the carillon began chiming for the first time. President Childs spoke to an all-school assembly to explain these changes to a concerned student body. In the fall of '66, Senior Court was halted by the administration, Mines moved into a new and enlarged athletics conference, the beer license was rejected, and the Young Republicans were active while the Young Democrats did nothing.

In '67–'68, the question of tradition versus long hair was first brought up in the *Oredigger*, and articles on student rights appeared frequently. It was in '67–'68 that a Task Force appointed by the Colorado Commission on Higher Education criticized Mines' stagnation during the past two decades. The Task Force encouraged Mines "to turn from what it regarded as a mediocre school into a top-quality institution." In the spring of '68, Theta Tau was placed on disciplinary probation by the administration for forcibly shaving 2 freshmen in the cafeteria. Pat Coyle became a write-in candidate for the Student Board of Publication [SBP], to give students who agreed with him a chance to voice their opinions. Coyle ran on a platform "opposed to traditions which violate the civil liberties of students," and he called for a reexamination of Mines' traditions in that light. The big news in '68–'69 was of course the ROTC petition, with over 600 students favoring a voluntary ROTC. The spring of '69 saw a junior elected to SBP for the first time, and a liberal elected as *Oredigger* editor.

I returned to Mines this semester from Germany to find work finally begun on the Graduate and Professional Center, to find new fraternity houses on West Campus Road. I see that students with long hair can now safely walk to their classes. I meet students who have new ideas and new attitudes, who care about things other than just the next GP quiz. And of course I meet many new students who, just like many of the old students, have no opinion and don't care.

Chapter 7: Still Achieving, Still Pursuing

Neal Harr '54 and Larry Miller '56 on the peak named for Mines.

5. Lower-division courses in geology reduced from ten to seven hours and presented as basic earth science courses.

6. Six hours of upper-division free-elective courses added, with the sole restriction that a student may not use them for courses in ROTC, physical training, or his own option.

These revisions indicated that Mines was making a genuine effort to bring its general education offerings into line with North Central's wishes. One member of the Curriculum Committee, according to Vanderwilt's 1958–59 Annual Progress Report, noted that, in addition to providing students the knowledge and understanding of the basic sciences and the fundamentals of engineering and the ability to apply them in their chosen fields, the school should stimulate their personal development by providing an intellectual environment in which they can fully realize their natural potentialities:

> The Mines graduate should have certain powers of thought. He should be able to reason cogently and to deal with the abstract as well as with the concrete and should have intellectual integrity. The Mines graduate should be able to express himself effectively both in speech and in writing. As a responsible member of a learned profession, he should have some knowledge of the economy and the place of his society in civilization. He should have moral integrity, accompanied by a sense of obligation to individuals, groups, and nations. As a well-developed human being, he should have some appreciation of the cultural—preferably an appreciation sharpened by an awareness of aesthetic criteria.

This statement no doubt provided grist for the profile of the Mines graduate that was to be formulated in the early 1980s. Though not revolutionary, these adjustments apparently satisfied North Central enough to welcome Mines back in April 1960.

II

The influx of both veteran and civilian students into the nation's colleges and universities following World War II impacted large and small institutions. Many smaller schools expanded their missions and increased their enrollments dramatically. Teachers colleges became full-fledged universities, and many liberal arts colleges expanded into additional professional areas with programs such as business and nursing. Some even added graduate programs. A number of large private universities such as

The new Gymnasium, completed in 1958 and named for Russell H. Volk in 1974, Mines' centennial year.

A view of the campus from the steps of Guggenheim.

Cincinnati, Louisville, Pittsburgh, and Temple became state supported. The exponential increase in knowledge, along with the increase in numbers of students completing undergraduate programs, pressured graduate schools to increase their enrollments. Many schools offering master's level programs, at the urging of prestige-hungry faculty members, rushed to add doctoral programs, even without sufficient resources to support them. Four basic arguments were put forward to justify graduate studies: they would bring in money from both industry and governmental agencies; they would attract strong faculty; they would strengthen undergraduate programs; and they would keep institutions on the cutting edge of knowledge. Counter arguments included cost in terms of monetary support and faculty time; weakening undergraduate instruction through the use of teaching assistants; and dissemination and duplication of effort among institutions.

Although Mines experienced a relatively long history of graduate education, it was its undergraduate program that had earned the school its national and international reputation. In an effort to strengthen the school's graduate programs, the Board of Trustees in 1953 established the position of Dean of the Graduate School and in 1954 recognized a graduate faculty. Graduate enrollment was sixty-three in the fall of 1954 and sixty-seven in the spring of 1955. By 1957–58, graduate enrollment had reached eighty-five for the first semester and ninety-one for the second. Geophysics had the highest enrollment, followed by Metallurgy. Of the 190 degrees granted that year, 15 were Master of Science and 4 were Doctor of Science. Respectable figures for a school the size of Mines, they provided a base for growth at the graduate level.

In 1957–58, among graduate fellowships awarded were twenty-six from Mines, one from the Mines Foundation, twelve from industry, and two from the National Science Foundation. Faculty research projects totaled $45,000 from the Colorado School of Mines Foundation, a twofold increase over that of the previous year. In addition, six industrial fellowships were granted. Not surprisingly, some faculty members and alumni were concerned that increased emphasis on graduate studies might weaken the undergraduate program. The general feeling, however, was that progress on the graduate front would bring a greater share in the largesse becoming available from a government anxious to surpass the Soviet Union in science and technology and would enable the school to maintain its international reputation.

During the following year, the Graduate Council, in an effort to address the need for combined graduate options of an interdisciplinary nature, arranged for graduate students to merge mineral engineering with basic science and mathematics. Offered provisionally for the 1959–60 academic year, this arrangement enabled students to enroll in the areas of chemistry/geology, chemistry/metallurgy, physics/mining, and physics/metallurgy. An exchange program with the University of Colorado permitted students from each institution to take courses at the other without additional tuition costs.

In prior years, many students who came to Mines for graduate study were required to make up work necessary to match the basic courses they would have received had they been undergraduates at the school. To ameliorate this situation, the Graduate Council strengthened the entrance requirements and testing procedures by establishing a qualifying examination covering chemistry, English, mathematics, and physics. Qualification in English was an important area for foreign students.

A ride through the Rockies in the Mines limo.

Dr. Orlo E. Childs, the eleventh president of Mines.

Undergraduate enrollment for 1957–58 was 1,127 for the fall and 1,027 for the spring. Geology options led the way with 337, followed by Metallurgy, Petroleum Production, Geophysics, and Mining in that order. Not surprisingly, these figures represent a steady gain in overall enrollment during the 1950s. Keeping pace with enrollment, the faculty numbers through the 1950s increased slightly each year. Physical facilities were also expanding. Arthur Lakes Library was completed in 1955 at a cost of $800,000 in state money. With three floors and five stack levels, it provided space for 150,000 volumes. Other features included a browsing lounge, four reading rooms, a map room, a large meeting room, seminar rooms, typing rooms, music-listening rooms, and a microfilm-reading room. The new library was welcomed for many reasons, one of which was that it freed space in Guggenheim, which had been home to the library since its construction in 1906. A much-needed dormitory, Randall Hall, was completed in 1957. Named for Bishop George Randall, the school's founder, it was financed by funds invested in government bonds starting in the fall of 1950. With the demolition of Old Chemistry, Hill Hall (Metallurgy) was completed in 1958 at a cost of

Paul Meyer Hall, completed in 1963, honors one of history's major mathematical minds.

$1.25 million of state funds. It was named for Nathaniel P. Hill, an original trustee of Mines, who, shortly after coming to Colorado in 1865, introduced the first effective process for smelting precious metals from complex sulfide ores.

Aside from the glitch in accreditation by the North Central Association, the 1950s had been good to Mines. Undergraduate enrollment was steady, the Graduate School was growing, the faculty was growing, and new buildings were turning Mines into a handsome, modern campus.

III

The decade of the 1960s was one of dramatic changes for America: the youthful president who was assassinated; the civil rights movement that resulted in significant steps toward racial equality; technological advances that broke the barriers of space; the impact of the computer on collecting and processing scientific data; and the war in Vietnam that wrenched the nation politically to an extent not seen since the Civil War. Changes in higher education were reflected in increased numbers of students and in the way those students perceived themselves both as individuals and as a part of a revolutionary culture that would move the nation in new directions.

In 1960, mineral engineering enrollment in the United States decreased by 7.6 percent. Mines, however, held fairly steady with a decrease of less than 1 percent. Moreover, according to the U.S. Office of Education, Mines led the nation in 1960 in undergraduate enrollment for geology, geophysics, and mining, and ranked third in both metallurgy and petroleum study at that level. Mines' undergraduate enrollment reached 959 and graduate enrollment reached 117, representing forty-eight states and thirty-two foreign countries. A total of 176 undergraduates and 23 graduates received Mines degrees in the 1961 commencement. All gained employment, went on to graduate study, or entered the armed forces.

The Horizon Plan was having its intended impact, both in terms of new buildings and in the addition of high-quality faculty. Vanderwilt was particularly pleased with the faculty, noting in his 1960–61 Annual Report that nearly "one-third of the faculty were involved in specialized studies and researches on their own, producing technical and professional papers or published articles totaling 124 separate pieces; the year's production was one-third more authors and twice as many papers and articles as in the previous year." Supported by both state and private funds, ninety-two faculty members attended 136 professional meetings. Perhaps the North Central Association's view, correct or not, that the

Students have always run for class!

Chapter 7: Still Achieving, Still Pursuing

The Ben Parker Student Union was an attractive addition to the campus.

Mines faculty was not overly impressive served as a positive stimulus. Either way, research was on its way to becoming a more significant factor in the Mines educational mission.

Faculty salary averages in 1960 ranged as follows: $10,042 for professors, $7,953 for associate professors, $6,848 for assistant professors, and $5,917 for instructors. While these salaries indicate an improvement, they were low for schools offering the Ph.D. Indeed, a number of large suburban high schools around the country paid teacher salaries as good as or better. The faculty numbered 26 professors, 28 associate professors, 37 assistant professors, and 16 instructors, distributed among the departments as follows: Geology, 18; Basic Engineering, 15; Chemistry, 14; English, 11; Mathematics, 10; Physics, 9; Metallurgical Engineering, 8; Mining Engineering, 6; Geophysical Engineering, 5; Physical Training, 5; Petroleum Refining Engineering, 4; and Petroleum Engineering, a mere 2.

Mines, however, still had no de jure tenure policy. Under what was generally referred to as de facto tenure, all contracts were renegotiated on a yearly basis. Basically, all associate and full professors were considered tenured, though the administration had considerable flexibility in negating de facto tenure appointments. Not until the late 1980s would

Mines have a tenure policy acceptable to the American Association of University Professors. Perhaps because the school was so close to the industries for which it produced engineers, it is not surprising that there was an industrial, top-down slant to some of its administrative operations.

Admission of undergraduate students was based on the pattern of course work and marks in secondary school; rank in graduating class; Scholastic Aptitude Test scores from the College Board Examination; any other test scores from secondary school; personality ratings; recommendation from principal or counselor; physical and social maturity; and motivation or desire to become an engineer. These items were all given essentially the same weight, with each applicant judged as an individual. The mean scores on the SAT for 1961 were 535 Verbal and 600 Mathematics. No applicants were admitted on a probationary status. Advanced standing was granted for course work completed elsewhere with a mark of C or better and an overall 2.5 grade-point average (2.0 for Colorado residents).

Freshmen from thirty-two states, one territory, and seven foreign countries enrolled at Mines for the 1961–62 year. Colorado with 38.9 percent of the total freshman enrollment increased its plurality over other sources. California, New York, Illinois, Texas, and Ohio followed. The increase of Colorado

The fireside lounge in the Ben Parker Student Union was a pleasant spot for relaxation.

View from the terrace of the Ben Parker Student Union.

students, according to Vanderwilt in his Annual Report of 1961–62, resulted from information spread across the state about broadened curricula and increased manpower needs, along with more intensive efforts by the Admissions Office. For 1962, the freshman class was all male. There were, however, three women enrolled at Mines, a faint hint of what the future held.

For several years, various faculty groups had been studying the expansion of the curriculum to offer new degrees in Chemistry, Mathematics, and Physics. During the 1961–62 academic year, the faculty recommended that such expansion be carried out. Degrees in Chemistry, Mathematics, and Physics would provide engineers of the minerals industry with much needed competence in those areas, as well as in related fields. A group of twenty-seven prominent alumni, chaired by Dr. James Boyd, director of the Bureau of Mines and former Mines administrator, were invited to visit the campus in January 1962 to examine the educational status of the school. According to Vanderwilt's annual report of that year, this group, after lengthy discussions and interviews, concluded that the expansion was "logical and necessary to better serve the State's educational needs and the technological requirements of the mineral industry and related fields of Colorado and the nation."

The Board of Trustees authorized Mines to offer these three degrees in April 1962, thus increasing the total number of degrees to nine. To placate those who feared that such a swing toward science would cloud Mines' mission of serving the mineral industries, these degrees were given hyphenated titles, e.g., Mineral Engineering-Physics. This decision proved a shrewd one for Mines as it pressed forward into a future quite different from the familiar present, just as would other decisions soon to be made regarding new degrees in Engineering (Civil, Electrical, and Mechanical) and Mineral Economics.

Vanderwilt's Annual Report 1962–63 coincided with his retirement from the presidency. He presented a short summary of his thirteen-year tenure as president, focusing on the positive aspects of enrollment, curriculum, research, buildings, alumni relations, and fundraising, all areas in which his leadership played a significant role. As might be expected, the theme of this last report was one of change:

> What then of the future? The changes of the past 13 years may seem unnecessarily emphasized. I have come to feel them to be normal and no greater than the changes that have occurred in past decades. What then of the future? So long as the nation grows and progresses, Colorado and the Colorado School of Mines must keep in step. Therefore changes are to be expected in the future as in the past.
>
> It is my belief that the future of the Colorado School of Mines is bright. Certainly Colorado and the nation, with growing demands for metals, minerals, and energy-producing materials of all kinds, will continue to need mineral engineers to study and develop the natural mineral resources to meet these demands.

While the academic side of Mines was developing new dimensions, the school's basketball teams of the early 1960s were making history as well. Although the Orediggers had not won a basketball championship since 1939–40, the Mines spirit was still alive. The 1961–62 team finished with a dismal

If an army travels on its stomach, college students study on theirs—as this view of the Union cafeteria shows.

Let's face it; the world is sometimes upside down.

9–15 season but won the Rocky Mountain Faculty Athletic Conference championship with a 9–7 conference record. Ironically, while Mines was losing its final conference game to Idaho State, Colorado State College was defeating Colorado College, the team that was tied with Mines for first place. Happily, the old expression of "couldn't win for losing" was proven wrong; and Mines was crowned the conference champion.

The championship was sweet, but the event that put Mines in *Sports Illustrated* (25 December 1961) bordered on the unbelievable. In the third game of the season at Spearfish, South Dakota, the Orediggers tangled with the Black Hills Teachers in a contest that took four overtimes to decide and in which Mines, with but ten players, was whistled for forty personal fouls. Near the end of the third overtime, with Mines trailing, the sixth Oredigger fouled out. Undaunted, the four remaining players tied the score, sending the game into the fourth overtime. Two more Miners were soon sent to the bench, leaving only Ken Ancell and Boyd Watkins to face five determined Teachers for the final two

The "M" climb always looks the same.

minutes. The Teachers took a three-point lead in the final seconds. Watkins, in one last gesture of defiance, drove the court for a basket. Final score: Black Hills Teachers 80, Mines 79. In the spring of 2003, this team was inducted into the Mines Sports Hall of Fame.

The 1962–63 team posted a 16–9 record, the first winning record since 1939–40. Because both the 1961–62 and 1962–63 teams lacked height, Coach Jim Darden cleverly utilized a control-style offense, sometimes referred to as a "hold 'n' hope" strategy. While not so short that they had to reach up to dribble, both teams averaged barely over six feet. Indeed, Darden towered over his entire squad. There was no time-of-possession rule. Control ball (with no shot clock) was a pleasure to watch, especially if your team was ahead, with its three- and five-man weaves that emphasized passing over shooting. At a time when collegiate teams were habitually scoring in the 80s or higher, Mines led the nation in defense in both these seasons, holding its opponents to an average of 52 points per game to its own 55.

IV

After a twelve-month national search, Dr. Orlo E. Childs assumed office as Mines' eleventh president on 1 July 1963. Selected from a pool of over fifty educational and industrial leaders, Childs came to Mines from the United States Geological Survey, where he directed research in marine geology and hydrology at Menlo Park, California. Childs received his undergraduate degree in 1935 and his master's degree two years later from the University of Utah. In 1945, he received his doctorate from the University of Michigan. He taught at Weber College in Utah, the University of Michigan, Colgate University, and the University of Wyoming. From 1949 to 1962, he served as projects director for the Denver office of Phillips Petroleum Company. Ben Parker, Board of Trustees president, described Childs as a man with "a distinguished record of professional service both to industry and education" and as "a vigorous young executive who typifies the school's excellence."

If Vanderwilt left his presidency on a note of change, so did Childs begin his. He titled his 1963–64 Annual Report "We Look Ahead." The report credited previous administrations for building "an outstanding physical plant, excellent teachers, and a tradition of service to the mineral industry that extends around the world" and emphasized that "the end of one era or of one program must be but the beginning of another. With increased pressures to

Every college has its youngest graduate. Here is Mines'.

educate more engineers while retaining the excellence that has been the hallmark of Mines, there is no alternative to growth." Perhaps recognizing the problem that Mines had experienced in satisfying both the North Central Association and the Engineers Council for Professional Development, Childs defended humanities:

> Preparation for engineering and science careers cannot be made solely in engineer- and science-oriented curricula but through the liberal arts and the other intellectual and spiritual disciplines [as well]. Mines must become a place where students will have freedom to seek, to explore, to find out for themselves, to experiment, to learn by error, to change their minds and their beliefs upon a more perfect revelation of the truth, and to hold fast to these beliefs without external molestation.

The long-range plan that Childs proposed featured a communications and humanities center with facilities that would allow for all-school assemblies, cultural activities, and professional meetings. The sketch of the proposed building looks much like the Green Center of today, a building that has achieved at least some of what Childs had in mind.

Mines' building spree begun in Vanderwilt's administration continued into that of Childs. In the last years of Vanderwilt's presidency, Chauvenet Hall had undergone considerable remodeling, ranging from the campus heating plant to a classroom building for Mining and Civil Engineering. The Assay Laboratory was converted to a laboratory for general research, especially in rock mechanics. Childs saw the completion of construction already planned or underway and proposed ideas of his own. Beauty, comfort, and utility were the characteristics sought in new buildings and were, for the most part, achieved. When Childs arrived, construction was scheduled for seventy-five new apartments for married students on state land but financed with federal

A view that never grows old: Guggenheim tower and the "M."

funds, a student union, and a science building to house physics. The apartments would accommodate one-third of the married students at Mines. Construction of the science building was financed with $775,000 from the State Building Mill Levy Fund and completed in 1963. A year later, it was named for Dr. Paul Meyer, the revered faculty member at Mines and practicing physician in Golden. The student union, known as the College Union, was completed in 1964 at a cost of $750,000, provided by a federal loan. Measuring 40,000 square feet, the union housed meeting rooms, a dining hall with seating for 500, a lounge area, a snack bar (the I-Club), a billiard room, offices, and a six-lane bowling alley. It was renamed the Ben H. Parker Student Center in 1970 in honor of the former Mines graduate, faculty member, president, and trustee. He served in the last position for nineteen years.

Coinciding with Childs' first year as president was a slight increase in enrollment that, in Mines' case at least, appeared to reverse the trend of declining

The Unknown (Student) Soldier
(source unknown)

In the old days (late Pleistocene), every male student at Mines was required to take two years of ROTC, unless they were veterans, as many were from the Korean conflict. A lot of guys had a bad attitude about ROTC and enjoyed making life difficult for the Army instructors. Several guys in the class of '61 decided to enroll a fictitious student, Dominic E. Perigo, in ROTC. They did so, and then turned in tests with Dominic's name—and lots of extemporaneous essay answers guaranteed to get Dominic very bad grades and an even worse reputation in ROTC circles. Soon the mere mention of Dominic, even in a stage whisper from the back of the class, was sure to get the instructor riled up.

The highlight of each ROTC spring was the INSPECTION. The student corps would march down to the intramural field and stand at attention forever while the general from Fort Carson and the PMS&T inspected us. It certainly was the highlight of the year for all of us really gung-ho types. The staff had drilled us endlessly and emphasized how critical it was that we look "like soldiers" and be all spit and polish. And also how important it was that we not lock our knees while at attention, like the guy that had fainted the previous year. As we marched down to the intramural field, muttering under our breath, the entire corps came to a halt at the gate to the field. It was locked!

This gate was never locked—but it was today. There was much scurrying around by the cadre. Some of the instructors were extremely upset at this SNAFU. Someone went running to find a maintenance man with a hacksaw while the student corps stood in rank with the general in a car behind us. After about thirty minutes, the man with the hacksaw appeared and removed the lock and chain. But by then, the brass had decided the INSPECTION would have to be canceled. The general's schedule was too tight to stick around. Boy, were we disappointed!

Oh, by the way, there was a tag on the padlock which said "Property of Dominic E. Perigo."

---

enrollments in engineering schools. Perhaps even more significant was a 40 percent increase in applications for the fall of 1963. At least some of this increase could be attributed to the new degrees in Chemistry, Mathematics, and Physics. Four women were enrolled for the 1962–63 academic year, and ten applied for the fall of 1963. Another factor no doubt contributing to a larger enrollment was the increase in financial aid. Scholarships held in 1962–63 comprised 38 Alumni Tuition Awards, 33 Special Award Tuition Scholarships, 9 Engineer's Day Scholarships, 20 Colorado Resident Tuition Scholarships, 52 Colorado Joint Honor Tuition Scholarships, 4 Price Foundation Scholarships, 4 National Merit Scholarships, 1 El Pomar Foundation Scholarship, 1 Gates Foundation Scholarship, 2 Boettcher Foundation Scholarships, and 3 Society of Exploration Geophysicists Scholarships. While these numbers may seem insignificant today, they represented an upward trend in financial aid benefiting both students and school.

From the fall of 1963, Mines' enrollment increased at a rapid pace, reaching a record 1,544 in the fall of 1966, showing 1,302 undergraduate students and

242 graduate students. Additional factors causing the rise included the crest of the postwar baby boom and an aggressive recruitment campaign by faculty, administrators, and alumni, who contacted prospective students and traveled to speak at high schools. The entering class of 1966 was strong academically as well as numerically, with 98 percent ranking in the upper third of their high school graduating class. SAT scores increased to 550 on the verbal and 640 on the math.

Early in his administration, Childs initiated consideration of a name change for the school (see article by James Robinson, MM January 1964). For some of the Mines family, merely the mention of a name change was heresy. Others considered it simply a logical step to be contemplated rationally. Opponents of such a change feared a diminishing of the school's educational reputation and recognized identity. Moreover, what would happen to old diplomas? Would they be reissued with a new name? Childs argued that the name of the school was misleading and ought to have "University" in its title, noting that Mines was probably the only

The Green Center, a vital academic and social center on the campus.

"school" in the nation granting doctoral degrees. Because of the vociferous pressure against any name change, the matter was abruptly dropped. Whether it will ever be reconsidered probably depends upon curricular and enrollment factors, financial support, and new waves of alumni.

V

Like Chauvenet, Childs considered Mines to be in a formative stage during his administration, as his 1963–68 Five-Year Report reflected:

> Thus, in retrospect, the years between 1963 and 1968 have been active and developmental, marking the progress of an emerging institution gradually and logically moving into a new and challenging position. The continuing studies, analyses, and planning that have marked this evolutionary process have been motivated by a deep concern for the rich tradition of Mines and by a vision of its future. These have been formative years, years in which the direction of the institution has been carefully planned to assure continuing, effective service to our society.

To be sure, Mines underwent a number of critical self-studies and outside reviews during the years in question, probably more than Childs really desired. Outside reviews consisted of Alumni Advisory Councils (1962, 1965, and 1968), a report by

Cecil H. and Ida Green, major donors for construction of the Green Center.

This E-Day event may well be the biggest ball game in existence.

Frederick Terman to the Colorado Commission on Higher Education (CCHE) (January 1967), an evaluation by the Engineers Council for Professional Development (March 1967), and a Task Force report from a group of distinguished experts appointed by CCHE (1967–68).

In his report, Terman, former Dean of Engineering at Stanford University, was critical of the status of science and engineering at the University of Colorado, Colorado State University, and Mines. With respect to the long-term prospectus, he recommended "an impartial, objective, unhurried, exhaustive examination." The accreditation evaluation by ECPD was basically complimentary, with easily correctable exceptions. In the meantime, the CCHE appointed a Task Force to carry out the examination recommended in the Terman Report. Chaired by Carl W. Borgmann, Advisor on Science and Technology at the Ford Foundation, the Task Force report was titled "Directions for the Colorado School of Mines," but was familiarly referred to as the Borgmann Report, which touched on numerous areas of the school's operation. The following are selected and edited findings from the Borgmann Report:

1. Mines has an excellent reputation at the undergraduate level in the fields of its specialties and is one of a few such institutions of higher education with graduating classes large enough to keep the per-student costs reasonable.

2. If Mines is to continue to educate for leadership in the mineral industry, its graduate programs must be materially strengthened. Mines, when compared with the national ratio of undergraduate/graduate degrees in its specialties, lags well behind the national average.

3. In academic ability, the Mines undergraduate student body appears to range from fair to good in comparison with those in other engineering colleges in the country. Its morale and esprit de corps are excellent.

4. The faculty, like the student body, is judged to be fair to good when considered on national standards. A comparison of faculty salaries in engineering at Mines with those at its sister Colorado institutions makes it clear that Mines lags at all levels except that of instructor.

5. The physical plant, with an exception or two, is judged to be adequate.

6. The level of support from sources other than student tuition and the state is low. Support by alumni is fair. Similarly, support by industry is modest. The amount of support of research and other activities by federal agencies is extremely low.

7. One weakening quality that appears to stand out at Mines is a combination of introversion and self-satisfaction. A majority (and sometimes all) of the trustees have been alumni. When outside advice is sought, in most cases a committee of alumni is assembled.

8. It appears clear that should Mines continue to concentrate its educational program heavily on the undergraduate level, the worldwide reputation Mines earned in the past will be shifted rapidly to other institutions.

In conclusion, the Task Force presented three alternatives regarding the future of Mines:

1. Curtailment. Abandon graduate education at the doctoral level and concentrate on undergraduate education.

2. No major change. Stay with a good undergraduate program and fair graduate program.

3. Development of national leadership. The Colorado School of Mines provides perhaps the best existing base in the United States on which to build one of the nation's leading technological institutions.

Like the "M," the ore cart pull to the state capitol on E-Day is symbolic of Mines.

New dormitories named for Jesse R. Morgan and Lester C. Thomas (1967) helped to ease the housing situation on the Mines campus.

The Task Force itself unanimously favored the third option, stating that "difficult as its achievement may be, there appears to be enough of a chance for real success to make the effort worthwhile." The CCHE accepted the third option on 26 April 1968, with a stated intention to support the efforts of the school and its friends to enlarge resources and to strengthen its programs. For a more detailed look at the Task Force's report, see Professor John Haun's excellent article "The Future of the Colorado School of Mines" (MM December 1968) or a copy of the report itself, available from the Office of Academic Affairs.

Another question evaluating agencies sometimes raised was Mines' extraordinarily large semester-hour graduation requirement. From a differing standpoint, the same question was often asked by alumni years before the 1940s: why students of later years needed considerably fewer credit hours for graduation. Before 1942, Mines followed a rather peculiar system that specified homework hours in addition to classroom hours, ballooning the graduation requirement to 448 credit hours, an average of 56 per semester and giving Mines bragging rights in quantity of work required. Most colleges or universities, however, expected two hours of preparation for every hour spent in class. Thus a three-hour course would add up to nine hours per week of study. Theoretically, in such a school, a student carrying seventeen hours would be spending fifty-one hours total per week on his courses. In that sense, Mines students were really doing just five hours more work than students at other institutions—hardly myth-making. Following reductions in the mid-1950s and in the late 1960s, the graduation requirement at present, depending upon the option, ranges from 135 to 148 semester hours, translating roughly into 18 to 19 hours per semester, not counting summer field camp. In contrast, most institutions require approximately 126 semester hours for graduation—or close to 16 hours per semester with no field camp. So, what does all this add up to? Just about whatever a particular school wishes it to. For an informative look at credit changes and degree offerings at Mines, see James Gary's article in MM February 1978.

At its February 1968 meeting, the third Alumni Advisory Council recommended significant degree changes for Mines:

1. Granting of Bachelor of Science degrees in Chemistry, Mathematics, and Physics and developing these programs through the doctoral level.

2. Continuing the science-engineering bachelor degrees in Chemistry, Mathematics, and Physics.

3. Eliminating the Professional Degree as an undergraduate degree.

4. Accepting the Task Force's third option for Mines' future.

5. Integrating more closely the activities of the Mines Research Foundation with the faculty.

6. Establishing advisory committees to departments not necessarily limited to alumni.

On 29 March, the Board of Trustees approved granting the Bachelor of Science degree in either Engineering or Science and the Master of Science degree in Chemistry, Mathematics, and Physics and the Master of Science degree in Mineral Economics.

The board also approved granting the Ph.D. degree in Engineering and Science. Again, as five years earlier, some alumni feared that these fundamental science-degree programs would eventually militate against the school's reputation in mineral engineering. Of particular concern was the dropping of the Professional Degree in Engineering, essentially a five-year degree that had its own strong tradition at Mines. In a compromise, the last year of the Professional Degree was moved to the graduate level with the equivalency of a master's degree. The basic argument in favor of the changes was simply meeting changing needs and hopefully attracting more and perhaps stronger students.

The above sweeping changes reflected what both internal and external groups thought necessary. Clearly, Mines was moving into areas beyond those laid out at its founding and in place for almost a century. One of the functions of CCHE has been to approve or disapprove programs that Colorado state institutions of higher education seek to put into their curricula. That Mines was able to gain approval for the above curricular changes was a testament to the united efforts of alumni, faculty, administration, and trustees. Again, for a comprehensive and readable summary of all that happened in 1968, see Haun's article in MM December 1968.

The above studies and reports show clearly how difficult it is for a school to meet the needs and expectations of its internal and external constituencies, particularly so when the Medusa of economics ever lurks unsatisfied in the background. Formative period, turning point, new direction, change—all were attempts to understand, explain, and foresee those times of redefinition. Near the end of his Five-Year Report, Childs highlighted Mines' efforts to ready itself for the future: modernizing the curriculum, strengthening the faculty, and emphasizing graduate education and research: "From these bases of self-analysis, the institution has reached a point of development which makes possible wholehearted acceptance of the long-range goals recommended by the Task Force." For Mines, the 1950s and 1960s were watersheds of curricular development and predictors of new demands and goals to be recognized and molded into reality.

CHAPTER 8

Mines at "100"

There are two kinds of doors: there are doors in space and doors in time. Doors in space you have to look for. Doors in time you have to wait for.

—Françoise Gilot

Influenced by such groups as the beat generation, peace activists, and flower children, college students of the late 1960s sought to realize their individual freedom in dress, hairstyles, music, dance, language, sex, and political views. Snake dances, freshman hazing, moonlight serenades, water fights, and panty raids gave way to more threatening kinds of entertainment, sparked by alcohol, marijuana, and LSD. The term "party school" came into vogue; and colleges and universities were ranked in that category, as they were in student SAT scores, faculty-student ratios, and the number of trees on campus. The wild card was the Vietnam War and the threat of male students being shipped to distant jungles to fight a war that many of them did not understand or accept. Larger campuses were often scenes of demonstrations, sit-ins, and acts of destruction as students and often nonstudents protested what they perceived as the "establishment." Seen as a symbol of a too-powerful military, the ROTC on many campuses became an easy target. When the Ohio National Guard troops, many of whom were in their late teens or early twenties and perhaps students themselves, fired on the predominantly student rioters who had burned the ROTC buildings at Kent State University in the spring of 1970, the terrible realization that "these are our children" wrenched American culture in ways that we may still not fully understand.

I

Although the burning and shooting at Kent State served as a climax, many schools in the late 1960s had been debating the place of ROTC in their curricula. Some dropped their units altogether, while others reduced their programs. One of the few still requiring the first two years of ROTC, Mines considered its program an essential component of the school's educational experience, providing both military and engineering training, winning many unit awards, and, most importantly, turning out hundreds of officers for the Army Corps of Engineers. Thus, there were both surprise and concern when a group of twenty-three students submitted a thirty-six-page brief recommending that ROTC at Mines be voluntary, as it was at other Colorado institutions. An editorial in the *Denver Post* of 17 February 1969 praised the group for not storming into the president's office with an ultimatum but instead "presenting the Board of

Opposite page: Mines marks one hundred years.

Dr. Guy T. McBride, the twelfth president of Mines.

Trustees with a thorough and brilliant report on the issue." The board, too, in a letter of 21 February 1969 to J. Douglas Pitts, president of the Associated Students, indicated its unanimous commendation on the quality of the brief and on the committee's mature presentation.

The board listed four alternatives to be considered:

1. The continuation of the compulsory ROTC program with appropriate modifications.

2. The continuation of an ROTC program structured much like the present program, but voluntary.

3. A voluntary program, sufficient as a base for obtaining a commission in the Army Corps of Engineers, but substantially different from the existing program.

4. Termination of the ROTC program entirely.

The board appointed a committee of three students, three faculty members, and an administrator to review the brief and to choose one of the alternatives. After a two-month study, this committee, in a close vote, recommended that the program should remain as it was. The recommendation was turned over to the Mines Curriculum Committee, which, after considerable deliberation, was unable to reach a majority decision. At its meeting of 9 May 1969, the board decided to continue the program as then constituted for the school year 1969–70. After further study and some restructuring of courses during that year, on 8 May 1970, the board passed a resolution requiring ROTC for the freshman year only. This requirement would remain in effect so long as there was sufficient voluntary enrollment to ensure a viable third- and fourth-year program. The board deemed 150 students the necessary number. Women, foreign students, those with military service, those Selective Service classed as conscientious objectors, and transfers classed as juniors or seniors would be excused from the requirement. The ROTC issue, however, was not fully resolved.

II

With Childs' resignation and departure to Texas Tech University as vice president of research at the end of the 1969–70 academic year, the Board of Trustees appointed Dr. Guy T. McBride president, a man with a highly successful career in education and industry. A native of Texas, he completed a Bachelor of Science degree in Chemical Engineering from the University of Texas and a Doctor of Science degree from the Massachusetts Institute of Technology in the same field. Following experience as a research

Caldwell Hall.

The utilization of a former fraternity house as a dormitory for women (Caldwell Hall) represented a major step in welcoming women to the Mines campus.

associate and instructor at MIT, McBride spent 1948 to 1958 at Rice University, serving as a professor in Chemical Engineering, director of Graduate Research, and an administrator in Student Qffairs. Prior to his arrival at Mines, McBride was vice president and general manager of the Phosphate Division of the Texas Gulf Sulphur Corporation. Besides chemical engineering, McBride had interests in ecology and environmentql control, having served on the Committee for Pollution and Environmental Control for the American Mining Congress. A scholar, teacher, researcher, and administrator in education and industry, McBride was well prepared to lead the school toward its centennial and beyond.

In the summer of 1970, McBride, in an early effort to gain perspective, requested each vice president to prepare a memorandum evaluating Mines' immediate past, present, and future. These were Truman Kuhn, vice president of Administrative Affairs; A. W. Schlechten, vice president of Academic Affairs; Gurnett Steinhauer, vice president of Business Affairs; and Anton G. Pegis, vice president of Development and Student Affairs. Not necessarily intended for anyone beyond McBride and the vice presidents, these memoranda collectively provide a broad and insightful picture of the school. McBride, in a handwritten note, complimented Pegis: "I have just read your excellent report for the third time. It's a great job from every angle, and I want to thank you for the effort you put forth." Pegis' memorandum concluded with a list of specific and perceptive recommendations, a selection of which follows:

- In light of the recent *Goals of Engineering Education* study, conducted by the American

My Memorable Mines Moment
(from Greg Staff '73)

The moment occurred, I believe, during Homecoming 1976. I was working for the school in the Admissions Office, having graduated in '73 and doing part-time grad work.

My fraternity, Beta Theta Pi, for which I was faculty advisor, had built an immense homecoming float, which was a fifteen-foot-high papier-mâché "Mine Shaft to Hell" constructed on a flatbed trailer. Completely surrounding the 40' flatbed trailer was the traditional skirt, a five-foot-wide edging made entirely of napkins "donated" by the surrounding fast-food establishments. The mine shaft had several moving parts—pulleys, ropes, etc.—that were kept moving by various and sundry undergrads inside the shaft pulling strings and pushing levers. Although I am probably biased, the float seemed at the time to be far and away the most intricate and finely decorated of them all, and don't forget that winning the float competition practically guaranteed winning the coveted Homecoming Trophy. Winning this trophy was the end that justified the means, the means being a copious quantity of Coors that was consumed during the float's construction.

As the float proceeded down Washington Avenue, driven by a college kid, don't forget, the fun was to begin. One of the props in the float was a minor explosion to be set off by a fuse traversing the back half of the trailer. The fuse then entered the mine shaft, lighting the charge and exploding a huge smoke bomb in front of the judges' stand at Foss Drug. And it worked.

BUT . . .

The fuse created sparks, naturally, which caught the interior of the mine shaft on fire, and this in turn caught the skirt on fire. These flames were aided and abetted by a panicked driver, who "floored the semi, madly honking its horn and accelerating crazily down the street." My vision is that of a speeding 40' long, 15' high "torch" screaming down Washington Avenue with guys diving off at every opportunity, as it cut a swath through high school bands, majorettes, and groups of small children—miraculously injuring no one seriously. The trailer's brakes melted down and locked up at the Clear Creek Bridge. The fire department was able to save the cab, but not the trailer. I guess the guys were trying to get the fire to the firehouse. What efficiency!

The Betas did not win the float competition that year, as they were disqualified for (unintentionally) causing a major disruption of homecoming activities. They did have the dubious honor of being responsible for a few new rules being instituted—no flames on floats, floats must have fire extinguishers, etc. I'm sure all who witnessed the Flaming Float of Beta Theta Pi still remember it well.

Society for Engineering Education, and the expressed requirement of industry, continue reviewing our degree structure and curricula.

- In our curriculum study, become more familiar with the quality of preparation our students are getting in high school.

- Conduct an inventory and evaluation of our courses in humanities and social sciences, with a view toward updating and upgrading course content.

- Regain the confidence of the faculty by reaffirming its role in the decision-making process.

- Clarify the functions, responsibilities, and authorities of each administrator.

- Create an atmosphere conducive to heightened visibility for the administration among the faculty, and vice versa.

- Reestablish a cooperative, friendly relationship with the Alumni Association Foundation.

- Implement a development program based upon sincere and deep involvement of our

1974 E-Day Queen Nancy Money.

entire constituency, from the Board of Trustees to related industries.

Focusing clearly on areas that needed attention, many of Pegis' recommendations became part of short-range and long-range planning. Though none of the four reports implied that Mines was misdirected, they all offered hints on avoiding possible pitfalls. Certainly a signpost not to be ignored was the establishment of an AAUP chapter on the campus in 1969, coupled with discussion of the need for a faculty senate. The ingredients for faculty discontent at any school are almost always present, simply waiting for something, or someone, to fire the still.

III

One of the first items of concern confronting McBride was the ROTC question. A bill introduced in the legislature provided that military training would be optional at the state's institutions of higher education. The bill was obviously directed toward Mines, the only school in the state requiring ROTC. McBride, in a four-page statement to the House Committee on Education, eloquently defended the role of ROTC, both past and present, in preparing engineering officers for the Army. One of his key points was the following:

Homecoming Queen (1974) Lynn Ryan and escort.

Centennial Homecoming parade.

It should be emphasized that the undergraduate program in military engineering [at Mines] is not apart from or simply "added to" the curriculum. Many aspects of military engineering are of both immediate and lasting value to mineral engineers, and mining expertise is of real value to the Corps of Engineers. The Mines curriculum has been planned to facilitate and to take advantage of this interaction to the point that Federal support of the ROTC program constitutes a subsidy to mineral engineering education.

Three Mines students, *Oredigger* editor George McFall, Student Council Parliamentarian Don Kammerzell, and freshman Dan Law also addressed the committee, giving their views on the quality of the military courses and why most students voting in a recent poll preferred that ROTC be voluntary. On 17 February, the Committee on Education voted to set the bill aside indefinitely.

In the spring of 1973, a case arose that again brought Mines to the attention of a national organization. The following outline of events is based on the report of the American Association of University Professors committee investigating the case (see *AAUP Bulletin* March 1973). Winton Laubach, an assistant professor of mathematics for eighteen years at Mines, was losing his sight from a hereditary eye disease. Because of concern that his teaching abilities would be hampered, Laubach was advised that his contract for 1968–69 would be his last. At this time, all faculty contracts at Mines were written for one year, though associate and full professors might receive de facto tenure, frequently viewed as essentially no tenure at all. Laubach, however, did receive a terminal contract for 1969–70.

Laubach requested a hearing before the Board of Trustees. Although no hearing was held, the board reviewed his case and concluded that Laubach should be given a contract for the 1970–71 academic year and that the institution should provide assistance in the form of projectors and other aids to compensate for his handicap. The AAUP committee recognized Laubach's concern that such assistance interfered with his classroom procedures but felt the school should have expended more effort to implement the wishes of the board. The committee's opinion was that Laubach, "unless he could undertake the difficult task of learning to teach as a totally blind person, would probably have had to retire from active classroom teaching within a fairly short period of time after his services were in fact terminated."

The committee's report, which did not dispute the school's motives in the case, questioned its procedures, noting that Laubach "was not afforded

Alumni musicians mark the Association's one-hundredth anniversary.

Chapter 8: Mines at "100"

Ski team in action.

the protections of due process which should precede termination of appointment for reasons of financial exigency or for medical reasons that were not adequately demonstrated." The committee also noted, "Academic tenure, as tenure is generally understood and accepted in American higher education, for all practical purposes remains nonexistent at the Colorado School of Mines." On these last two points, the AAUP censured Mines, which remained on the censure list until 1992.

IV

Despite the continuing conflict in Vietnam, the United States and the Soviet Union had, by the 1970s, arrived at a modus vivendi that, despite clandestine games of espionage and brinksmanship, enabled them to fashion a semblance of order in the world. At least a program was not necessary to recognize the players. Even though the unusual combination of inflation and unemployment gave pause to both economists and consumers, America felt relatively secure as a superpower. Following the Yom Kippur War, however, Arab states put into place an oil embargo that played havoc with American motorists and industries. The end of American self-sufficiency was becoming a reality, as McBride indicated in an address to the Newcomen Society in North America on 10 November 1973 in honor of Mines' upcoming centennial year:

> For these past many years, we have been living on the pay dirt; from here on in, it will not be so easy, for we have gone far in depleting the easy gold, the accessible oil, and other fossil fuels. We are having to turn more and more to remote and hostile environments for our resources, environments that will place enormous demands on all our engineering skills. . . . But we have not yet been able to make full use of these resources because we have not possessed the working knowledge to do so while at the same time protecting the environment—a necessity of which we have, to our credit, become aware for the first time in material history.

Incorporating ecological concerns to the process of extracting minerals from the earth was not easy for Mines or for the mineral industries, as McBride recognized: "Our thinking has been dominated by the 'bonanza' concept of wealth. Find it. Dig it out, spend it, and get out." However, "this thinking is passé for the future. We cannot depend upon it, even if it were possible that somewhere in the country there remained vast undiscovered wealth."

No snow and no skis, but the 1976 ski team is ready.

167

Excerpt from the Diary of a 1975 Mines Coed (Alpha Delta Sorority)
(from Ramona Heikel '79)

When all the planned orientation activities of the opening of the fall term were finished, we found ourselves creating some of our own original kind of fun: "raids." Whenever we snuck into a fraternity house unseen and did something that would get someone riled, that was a raid. One night early in September, while most of the campus slept, we stole the Beta Theta Pi's volleyball net and "Limburger-cheesed" the Kappa Sigmas and Betas. We also stole the ATO park bench and left it on the lawn of the Jefferson County Courthouse. They found it and put it back on their front lawn, but two hours later, we took it again. We hoped that all of this would liven things and that the guys would notice that there was now a sorority house on campus.

A week later, the Sigma Phi Epsilon guys found the fuse box to our house and shut off all our lights. Then they pulled a panty raid, roaming through our rooms wearing stocking caps over their faces and carrying flashlights. They rifled through our clothing and displayed their booty on the fraternity balcony the next day. To retrieve our personal items, each had to claim her own intimate wear in person.

Two days later at 4:00 a.m., we raided six frat houses (they were all close together), loosened the lids on the salt and pepper shakers and took anything that looked like it would be missed: composites, basketballs, a football (which I still have), the Sig Ep hippopotamus trash can, ATO trophies, and someone's volleyball pole. Many of us bought locks for our rooms that day, in anticipation.

The night after our raid on the fraternity houses, the Sigma Nus raided. They kidnapped a group of us, "creeked us," and let us go. As we were walking home in wet clothes from the creek, another group of Sigma Nus met us and took us back to the creek to finish the job properly (our hair wasn't wet yet). But we had to go to the Sigma Nu house to rescue our friends, and when we did, we got showered. Then they treated us each to a pop.

As we were walking home from the Sigma Nus, a truckload of SAEs kidnapped us, took us to their house, and shoved us into "the pit." I believe it was a storeroom of some sort. With kicking, hair pulling, and biting, we attempted to escape but were recaptured. After we were tied up, we were covered with shaving cream, then hosed off, and, in their kindness, were given blankets. I almost escaped after that but I couldn't find the front door in time. We rode home on the hood of someone's car and found that a party had broken out at our place.

But we didn't stay home long. The Betas had been so quiet that we figured they needed some provocation, so we walked to their house and took some things, and got caught and showered. Some of us got wrapped up in computer tape. I escaped and found the fuse box and turned off all

(continued on next page)

Whatever its causes, the oil crisis served as a backdrop to Mines' centennial year and underscored the school's importance to the mineral industries. As enrollment accelerated in the following years, so too did funds and resources for energy exploration.

In addition to celebrating its one-hundredth birthday in 1974, Mines also scheduled a North Central Association evaluation. The evaluating team met with all major administrators, all department heads, and most of the faculty. Noted strengths included the following:

- Strong esprit de corps on campus and among the alumni.

- Sustained enrollments counter to national downward trend.

- Attention to instruction in rigorous professional programs providing excellent experience in science and technological areas.

- A few departments such as geophysical engineering and metallurgical engineering with some outstanding faculty and very good research activity.

Although Mines retained its accreditation, North Central identified thirteen concerns—four of which appear below:

- School is no longer as well regarded in state as it once was.

- Vision of the future as developed by the trustees and administration unclear.

---

the lights. All of us ran out. Two of us ran straight over to try to stir things up at the Kappa Sigma house because they seemed to be interested only in studying. They, in return, pulled a panty raid on us the following night.

———

By Wednesday, October 1, it had been two long weeks without excitement, so a few of us went up at night to the ATO's and spray-painted their green park bench a bright fluorescent pink. That was a pretty major alteration, and we hoped they wouldn't be angry, but they said they loved it, except that we had missed a few spots.

Another time, the SAEs got hold of about six or seven live chickens and let them loose in our house. There were females screaming and laughing all over the house. Then two of us rounded up all of the chickens and put them into a large box, let them go in the SAE's living room and left without running into anyone. This was all in broad daylight for a change.

To break some of the tension during finals week, several of us "streaked" down frat row. Although it was quite cold, we wore only bikinis and ran in front of car lights so we could be seen, honking all the way down Prospect Drive. We decided to make a second pass because of a lack of response. Since I was cold, I rode in the car this time. Lucky for me because this time the streakers got enough attention. They got carried away, literally, to various frat houses.

As I remember these events during my first year at Mines, I am amazed at how little sleep I needed to function. And I also noticed that, as time went on and the less fun I had, the lower my grade-point average dropped. I'm just glad I was in the right place at the right time to be a part of the action!

Alpha Gamma Delta house.

- Policies and procedures relative to faculty hiring, firing, and promotion lacking in faculty involvement.

- Beyond economics, no apparent consideration of the relation between humanities and social sciences to the primary programs in engineering and applied science.

In their discussion of the North Central report, members of the Board of Trustees noted that a faculty poll showed majority agreement with the concerns noted in the report.

Given the continuum of time, into which succeeding years blend, the one-hundredth year of a school's existence has more symbolic than practical significance. Age and change, however, can bring painful concern for those who have a fixed idea of their school's identity and its function. Others welcome any change, even if it means grasping every new idea coming along, a tendency of which Sir Walter Scott warned: "But patience, cousin, and shuffle the cards, till our hand is a stronger one."

One shuffle of the cards that added flexibility to the Mines curriculum was the Basic Engineering Department's successful request to CCHE for a Bachelor of Science in Mineral Engineering that would allow students to plan with an advisor various course combinations from both the Basic Engineering Department and the traditional degree-granting mineral-engineering departments. The *Oredigger* of 5 February 1974 reported that the goal of the new degree was "to produce a graduate capable of handling a wide variety of engineering problems, as opposed to the more scientific specialties of the existing options." This new degree allowed Mines to expand its mission without distorting its base, a strategy that the school would continue to practice under McBride's leadership.

All of Mines' outside evaluation reports stressed the need for increased faculty research to strengthen its position as a leading institution in mineral engineering. An in-house report by John D. Haun, professor of geology; George H. Kennedy, associate professor of chemistry; and M. A. Wiltse, assistant professor of geology titled "Faculty Productivity, Quality and Financial Support at CSM," originally presented to CCHE in 1972, appeared in MM April 1974. The report, based on eighty completed questionnaires from Mines faculty members in the engineering and science departments, showed that faculty members spent an average of fifty-four hours per week in Mines-related activities, with 56 percent of that time devoted to twelve credit hours of course work. Student counseling, research, and professional and academic development occupied 21 percent of their time. The remaining 23 percent was spent on a wide variety of school activities. The report emphasized that "quality of teaching is as important as the numerical measures of faculty

Beta Theta Pi house.

Mickey Mouse Visits Mines

Many colleges and universities of the late 1960s and early 1970s faced the question of freedom of the press as it applied to their campus newspapers. Thus it was that Mines students working on the *Oredigger* during that period felt they should have the right to decide what was printed in it. The administration, on the other hand, believed it necessary to have some control over what appeared in the paper, lest the school's image be tarnished. Moreover, the school provided the paper with money and space.

One avenue for testing the question lay in both the nature and the language of articles appearing in the paper. Thus, when President Orlo Childs felt that some of the articles appearing in the *Oredigger* went beyond the bounds of decency in content and language, he called for a Board of Publications review of the paper's procedures. Open hearings marked by lively debates among students, faculty, and administrators followed. Childs verbally placed Editor Jack Yench on probation with regard to his status as editor of the paper, warning him that any more disciplinary violations on his part would result in dismissal from school.

Yench forced the issue when he published a College Press Service story verbatim in which a Canadian college president used expletives in telling the editors to clean up the campus newspaper. At a quickly convened Student Council meeting, Yench was removed from his editorship. When the entire editorial staff resigned in protest, the Student Council met again and voided their earlier meeting, citing improper notification. Yench, however, resigned his position as editor to publish the *Technocrat*, an independent campus paper.

One of several seniors with some credit hours to be completed in the summer, Yench was permitted to participate in the 1971 Spring Commencement. Foregoing the standard mortarboard (but not the tassel), Yench appeared with a set of Mickey Mouse ears on his head. Despite warnings that he was in violation of school protocol, Yench proceeded onto the stage and announced his own name into the microphone when the Dean of Students did not. The penalty was swift and severe: dismissal without a chance to complete his degree. Following an unsuccessful due process suit filed in the United States Court of Appeals, Yench left Golden to complete his education elsewhere and went on to a successful engineering career—with, as he says now, only positive recollections of his days at Mines.

In 1981, a group of seniors tried a different version of "Mickey Mouse Ears" when they designed and draped an elaborate set over the belfry of Guggenheim Hall and played the Mickey Mouse Club marching song over a loudspeaker. There is no record of any punishment. Whether it be dynamite blasts, Senior Day escapades, Mickey Mouse ears, or performance in the classroom and in careers, Mines students generally reflect intellect, imagination, and risk taking—even if not always to their advantage.

Sigma Nu house.

productivity" and that Mines faculty members were "directly involved in supervising laboratory exercises, recitation sections, and field problems," duties that in many universities are carried out by graduate students. In summary, Haun, Kennedy, and Wiltse asked that CCHE and the state legislature invest "the funds necessary to increase the numbers of faculty and support personnel to a level at which we may proceed realistically to Option 3 of the Task Force Report, and the resultant increased benefits to the State of Colorado."

V

In the first Presidential Report of Mines' second century, McBride focused on change, noting that "in institutions as in persons, change occurs rapidly at some stages of development and much less rapidly at others." McBride evidently saw "doors in time" that would open for Mines and believed that trustees, administrators, faculty, students, and alumni shared his view. Enrollment had increased from 1,687 in 1973 to 2,330 in 1976, and reflected a noticeable change in composition. In the fall of 1974, the 224 women enrolled ranked Mines among the higher percentages in the nation's engineering schools. The average entering freshman graduated in the top 12 percent of his or her class and averaged 535 (verbal) and 635 (math) on the SAT. One downside regarding enrollment was the decrease in out-of-state students, causing concern not only for student diversity, but also for the extra tuition that out-of-state students brought. Tuition for nonresidents at Mines was among the highest of state-supported institutions.

Even with the increase in enrollment, McBride underscored in his report that "the pattern of legislative appropriated support has been and remains inadequate to support quality engineering education at the school." Criticism leveled by the Task Force and by the North Central Association that the state was not supporting Mines at the level necessary for it to reach its potential as a leading institution in mineral-engineering education was still operative. The conclusion was simple enough: non-state resources must be developed to supply an increased portion of the school's capital and operating requirements. Following the Board of Trustees' approval of a table of needs, Mines announced a fund-raising campaign in February 1977, the Resource Fund, which intended to raise $63 million over a ten-year period. With William K. Coors as chair, the successful campaign received broad support from the community generally and from Governor Richard D. Lamb and the *Denver Post* specifically.

McBride, in 1977, asked Dr. Frank Hadsell to form a committee to create a document describing the educational goals of Mines. Dr. Rex Bull, the Committee's Executive Secretary, summarized the

Kappa Sigma house.

Sigma Alpha Epsilon house.

extended deliberations in a ten-page document listing six general attributes Mines graduates should exhibit. Later, these attributes, known as the Future Graduate Profile, became a key goal of a Mines education:

1. The graduate must have undisputed technical competence in some field related to minerals and/or energy.

2. He must have the ability to communicate his thoughts orally, in writing, and graphically, both to his fellow professionals and to the layman.

3. He should have been inspired and should have the background and desire to be able to continue to learn independently.

4. His experiences at Mines should have stimulated his natural curiosity so that enquiry, analysis, and syntheses based upon fundamentals should be a natural response to any unfamiliar situation.

5. He should have breadth to be able to appreciate the problems of professions allied to his own. He should realize clearly the potential impacts of his professional actions upon the political, social, economic, and natural environments in which he practices.

6. He should have, as any professional, high standards of integrity and self-discipline

and a positive attitude to the professional responsibilities of his job.

The report recommended more term papers and seminars, more emphasis on theory and practice, and more upper-level humanities and environmental electives. Probably few schools have been so concerned with the characteristics of their graduates and the curricula they study as has Mines. Bull added a poetic/musical touch to characteristics of the Future Graduate (first performed at McBride's retirement dinner and roast):

THE FUTURE GRADUATE

I am the future graduate,
I have a profile that's first rate;
I'm broad, I'm deep, I'm thick, I'm round,
And seldom can a fault be found.
 Oh boy, can I communicate!
 My ethics are immaculate,
 And as for learning—
 So much will be known to me,
 Politics, philosophy;
 I'll be perfect, I'll be perfect;
Some day I will graduate,
Maybe nineteen ninety-eight,
And go study law at D.U.

An external view of a Mines education, titled "The School of Hard Rocks," appeared in *Newsweek*. The article's simplistic approach cut through to the reality that the school itself had worked to create

Sigma Phi Epsilon house.

Without a house, the Barbs are an active group in campus affairs.

and that the students had come to expect: "On this campus there is little time for play. Students grind away most of their waking hours in lecture halls, labs, and libraries, or crawling through an ancient mine that serves as an underground classroom, or roaming the Southwest on contract fieldwork for the government or industry" (2 November 1981). Pointing to a heavily prescribed curriculum offering few liberal arts courses, the article noted that some prospective employers see Mines graduates as too narrow for modern business, particularly for top-management positions. The "Graduate Profile" mentioned earlier was seen as a corrective to the "cookbook approach" to problem solving.

The article emphasized that while the "sink-or-swim" atmosphere on the campus caused some students to wilt under the pressure, it also brought them closer together. Robert Burford '44 (later Speaker of the House of Representatives in the Colorado Legislature and head of the United States Bureau of Land Management) was quoted as comparing Mines with the Marine Corps in that respect. Did it pay off? The article indicated that it did: "[Mines'] prized graduates can be found running open-pit phosphate mines in North Carolina, hunting natural gas in Peru, designing oil refineries in Indonesia, and ruling the executive suites of energy and mining firms." Still, as McBride stressed, complacency is not a Mines byword and the school, to survive, would have to have a curriculum that would produce more creative engineers and better decision makers.

In 1979–80, through the efforts of McBride and the Board of Trustees, the legislature adopted the Mines Plan, an arrangement giving the Board of Trustees control of the school's budget and holding them accountable for performance as opposed to line-item accounting. The trustees also gained the power to adjust tuition as necessary, allowing for easier short- and long-range planning. In November 1982, the board approved a Master Plan (1982–88). This 120-page compendium of facts, figures, reviews, and projections presented an exhaustive view of Mines' present status and future direction. Written at a time when the economy was slowing, particularly with regard to metals and oil, and demand for engineers was declining, much of the plan unfortunately became irrelevant until the

1990s. The exercise, however, did identify possible options and prospects.

George R. Brown Hall (Mining and Engineering) was completed in 1980 and dedicated to honor Brown '22 in April of 1981. Brown and his brother Herman developed Brown and Root, Inc. from a small road-building company into one of the world's largest construction conglomerates. Brown Hall was built and equipped with $5.5 million of private donations, with $4.4 million coming from the Brown Foundation.

VI

Toward the end of McBride's presidential tenure, a number of Mines administrators, faculty, trustees, and alumni compiled a document measuring the progress of the school during his fourteen-year leadership to be used as the reference point for the Annual Trustees Conference of 1984 at Keystone. Titled *Benchmark*, in addition to covering McBride's 1983–84 President's Report, the booklet was "to provide an overview of McBride's administration—the fourth longest tenure of any Mines president—and to serve as a transition into a new administration. Adding valuable background analysis were the informative and frank interviews of trustees, administrators, faculty, alumni, students, and industry and government officials. Overall, *Benchmark* presents a clear picture of a school moving through a period of challenge and growth.

ROTC Headquarters.

What Mines students do best—take a test.

In the President's Statement, McBride listed three major decisions that, in his view, were crucial during his presidency (edited):

1. The decision to double the undergraduate-graduate enrollment from approximately 1,700 to a limit of 3,000.

2. The decision to seek major non-state tax funding to become a "state-assisted" rather than a "state-funded" institution.

3. The decision to alter the undergraduate curriculum to produce bachelor-degree mineral engineers.

He did not list emphases on graduate research and education because they had not yet been implemented to the same degree as the three curricular and funding decisions. He cautioned that decisions do not always solve problems, and that they may sometimes cause them. He then observed that a passing administration should not be judged solely by problems solved, but also by those left behind. Of these latter, he mentioned the following (edited):

1. Securing the funding base.

2. Stabilizing the total enrollment and the individual department enrollments.

3. Harmonizing the interrelated efforts of the school.

Freshmen participating in their songfest.

4. Improving faculty development and faculty evaluation procedures.

5. Creating a humanities program to be an essential part of undergraduate education.

6. Changing building space to match current and future personnel and program status.

McBride closed his statement with a word of advice that all collegiate presidents would do well to remember: "Each member of the Mines community stands on the shoulders of those who preceded him, and that which has been accomplished by any of us

Thespians and Authors

On a fall day in 1971, an announcement appeared on the door of Assistant Professor of English John Hogan inviting students interested in drama to come to a meeting. Fifteen came, and the Mines Little Theater was born. The first endeavor, a one-act play, *Adaptation* by Elaine May, was staged in the round in the I-Club under normal ceiling lights before a curious audience of fifty, who paid fifty cents admission. From this small start, Mines Little Theater quickly became a popular extracurricular activity. For five years, Hogan was the primary producer and director of two plays a year. Attendance grew to as many as three hundred jammed into the I-Club. The highlight of these years came during Mines' centennial year 1974, when the melodrama *She Was Only a Farmer's Daughter* was presented in Bunker Auditorium before an audience of nine hundred. Mines Little Theater had arrived and was ready for challenges with productions running the gamut from light to serious, e.g., *Fantasticks* and *One Flew Over the Cuckoo's Nest*.

When Hogan gave up his side career of producing and directing plays, he organized the Mines Little Theater Council, made up of students and faculty members. At this time, Dr. Steve Daniel of the Chemistry Department and his wife Linda became active as directors and performers in many of the productions. By this time in 1978, performances had moved to the large Chemistry lecture hall and would eventually move to Petroleum and Metals halls in the Green Center. For Daniel, the whole operation was "fun theater," aimed at getting students to participate and to attend. For his part, Daniel likes to portray villains, because, as he says, he can do "all sorts of mean things that I cannot do in the classroom." Roles he has played include Sugar, the devil in *Damn Yankees*, and Big Jewel in *Guys and Dolls*. The Daniels have also been active with the Mines chorus for many years.

Following his activities in drama, Hogan founded *High Grade*, a student literary magazine, in 1978. Each semester, this magazine publishes poetry and short stories, as well as student drawings and photographs, and has served well as another creative outlet for Mines students. Hogan served as advisor until 2001.

The Society of Women Engineers plays an important support role for female students.

and by all of us is less the result of individual or group effort and ability than it is in part a reflection of history and in part the resultant of complex external and internal factors beyond our complete understanding and effective control."

The following edited comments from interviewees, though representing only a fraction of material collected, offer some pertinent insights:

- It is crucial for the alumni association and the school to establish a pattern of rapport and interaction to strengthen the school's development.

- McBride brought a needed sense of direction at a time when the school was in danger of being absorbed by the University of Colorado.

- There needs to be more communication among departments. The school's program cannot be cohesive unless the prima donnas are talking to each other.

- The real value of Mines is that it teaches one how to work hard. It is impossible to get through Mines without appreciating the work ethic.

- The school could do better in the area of financial support. Mines has been handed a pair of loaded dice and is playing the nickel slots.

- Mines, similar to a private institution, is extremely fortunate with respect to autonomy. It is one institution, one student body, one faculty, one governing board, one president, one campus.

At McBride's last commencement as president in the spring of 1984, the Board of Trustees surprised him with an honorary doctorate.

VII

Besides evaluation reports by external agencies, arguments about ROTC, and AAUP censuring, faculty members were teaching and researching; students were going to class and graduating; athletic teams were competing; and fraternities and other groups were socializing. Perhaps of more concern to alumni of an "old" Mines was that some of their traditions were giving way to changes imposed by students of a "new" Mines. Freshman Orientation Week replaced Agitation Week and because of legal implications was made voluntary. To the tug-of-war between the freshmen and the sophomores, a pushball contest between the two classes was added. Freshmen participating most enthusiastically in the various activities would receive awards at Senior Court. A Bar-B-Q (not the wrinkling variety) would be held for freshmen that helped paint the "M." Surviving activities included the miner's cap worn by freshmen, the midnight songfest, and the civic project. Charlie McNeil, student body president, explained the philosophy behind the changes:

> With this year's orientation program, we're doing away with all harassment. We're

The Blue Key views the world from Mines Peak.

The Mines Student Council (1980).

providing the essential knowledge of the tradition and spirit of a Colorado School of Mines engineer. Since there is no harassment from the upperclassmen, the freshmen need something to pull them together to form a united class. That's why we're putting more contests in orientation. The program is strictly to benefit the incoming freshmen and all are strongly encouraged to participate. (*Oredigger* 16 September 1970)

Obviously, the interests that united students in the first half of the twentieth century were not operative in the second: World War II, the Korean War, the assassinations of John and Robert Kennedy and Martin Luther King, the cold war, TV, computers—all had a marked impact on how Americans, young and old, thought and behaved. At Mines, such activities as cruising the campus with clothes on backwards were no longer accepted. The challenge now was to keep alive the spirit that had inspired and sustained Mines for almost a century. In the *Oredigger* of 16 September 1970, Chuck Hahn and Dan Swartz reduced the entering class to two types: (1) the typical student who cares only for engineering and could attend any area engineering college and (2) the student who not only will strive for a top engineering education, but will follow ninety-six years of tradition to become a Miner with a genuine interest in the school. Interest, pride, and unity, according to Hahn and Swartz, are the factors that make the Mines student body unique. Such a view provides a clear and simple justification for tradition. When the stands at Brooks Field erupt with "The Mining Engineer," the moment is no less significant or inspiring than when 107,000 Penn State football fans shout "We are Penn State" or when Yale students wave their white handkerchiefs during the last lines of "Bright College Years" in the Yale Bowl. Implied perhaps in what Hahn and Swartz wrote thirty-four years ago is that defining the Mines spirit at a particular time is not so important as living it all the time. And that, of course, is left to all students and alumni who believe in it.

During its first one hundred years, Mines graduated thirty-seven women, hardly enough to make an impact on male domination. The steady increase in women students through the 1960s, especially as they began living on campus, brought a new element of sophistication that impacted the school's traditions. The former dean of students' house was converted in 1964 into the first dormitory for women and accommodated six women. Four years later, the school purchased the old Kappa Sigma fraternity house and renovated it for another women's dormitory, Caldwell Hall. *A Century of Women at Mines* notes that women, though still few in number, were creating their own traditions. Joan Bacon '66, for example, was the first woman to wear a senior Stetson. From 1966 to the end of the decade, graduating women received a red rose along with their diplomas. Barbara Fisk, the first woman faculty member at Mines, came from the University of Colorado two days a week during the fall of 1968 to teach a course in the appreciation of the fine arts.

While some of Mines' oldest traditions were going the way of the slide rule, interest in athletics remained strong. True, Mines was not competing with many of its earlier rivals, at least in the major sports of football and basketball; and the interest of the student body may not have been so focused as it was when the school was smaller and freshmen were required to attend football practice to support the gridders. The drama of kidnapping mascots and shaving the heads of those who attempted to deface Mines property was but a memory. Yes, Mines teams lost more often than they won. Nevertheless, the Mines spirit was exhibited on the field or court

Cheerleaders reaching high.

of play as vibrantly as ever. Improved facilities, more players, more coaches, and more scholarships—all were important factors in the high percentage of Mines students participating in varsity athletics. Some schools at the time may have fielded more teams than Mines but not many—and the same is true today. Athletics at Mines has never been motivated by money; and while "student-athlete" may be an oxymoron in much of the world of intercollegiate athletics, at Mines it is fact.

An article by James Sankovitz in *Coach and Athlete* of January 1960 illustrates the importance not only of intercollegiate athletics at Mines, but also the emphasis on physical fitness for the entire student body. Sankovitz commented that, even with the school's rugged academics, nearly 90 percent of the student body participated in sports: 600 in physical fitness courses, 175 on intercollegiate teams, and 200 more on intramural teams. He also noted that ROTC required several hours per week of physical drill for its cadets and that clubs such as flying, skydiving, mountain climbing, and gymnastics were available for students looking for additional competition.

The Mines student body at the time of Sankovitz's article, however, was still primarily all male, with no organized athletic program for women beyond bowling, cheerleading, or shooting. This situation changed during the 1970s, when the NCAA Division II sports of basketball, golf, skiing, swimming, track, and volleyball were added for

It's not a prison break. It's the Mines Rugby Club in action.

In 1981, a group of seniors designed and draped over the Guggenheim belfry an elaborate set of Mickey Mouse ears, accompanied by the Mickey Mouse marching song played over a loudspeaker system that carried across the campus.

Just as athletic competition has been an indispensable part of life at American colleges and universities, so too have been fraternities and sororities. When five young men at the College of William and Mary founded Phi Beta Kappa in 1776, they had no idea of the tradition they were starting. When Phi Beta Kappa shed its secrecy to become an honor society in 1826, the fraternity system was well on its way to becoming a significant force in American higher education. At Mines' founding, fraternities and sororities, to a lesser extent, were widely represented on American campuses.

Fraternities early became a major part in the Mines scene, providing fellowship and accommodations for almost half the men enrolled. The accommodations, moreover, were often in houses that resembled mansions. The current president's residence, for example, once housed the Sigma Nu fraternity. In 1964, the "mansion era" came to an end when county government and Mines necessitated acquisition of land occupied by several fraternities. As compensation, seven building sites on the west side of the campus were made available to fraternities at $5,000 a site. Five fraternities took advantage of this offer and built new houses to accommodate thirty-six to forty-two students. Without the impressive exteriors of some of their predecessors, their modern facilities and lower maintenance costs were plus factors.

women. According to *A Century of Women at Mines*, four women received athletic scholarships in 1975: Vera Fowler, ski team; Colleen Lynch, rifle team; Eloise Montoya, men's golf team; and Leslie Puttuck, swim team. Basketball and volleyball became intercollegiate sports for women in 1976. Although the first basketball team saw just a single victory against eleven losses, by 1979, when the Rocky Mountain Athletic Conference added women's sports, Mines women were competitive. Led by all-conference Sandy Wrobel, they won the East Division title in 1979 and finished second and third the following two years. The volleyball team won the RMAC title in 1982. Swimming competition began in 1981 in the Inter-Mountain Swimming League, followed by track in 1982 and softball in 1983 in the RMAC. A truly outstanding track performer, Sarah Thaler won the RMAC discus crown four times and the shot put three times. While the men profited from a one-hundred-year start, the women were making great strides.

Seniors celebrating their Mickey Mouse deed.

Finances have generally been a constant problem for many national fraternities and sororities. Occasionally, when a house was not filled to capacity and nonresident, dues-paying members were not sufficient to cover costs, alumni were called on for aid, just as they are today. After all, the secret rituals, passwords, handshakes, and initiatory pledges make up the bond of permanent fellowship regardless of age discrepancies—so the pledge manuals say. Additional activities created fellowship: beer parties, dances, Homecoming floats and house decorations, intramural athletic contests, hell (help) weeks, songfests, rush weeks, locking the actives out of the house for a night, and "pinning" your sweetheart. Like many secret societies or country clubs that select their members, collegiate fraternities and sororities on some campuses

Mickey Mouse would have been proud.

The Energy and Minerals Field Institute

An arm of Mines' Special Programs and Continuing Education (SPACE), the Energy and Minerals Field Institute (EMFI) has, since its founding in 1978, played a significant role in familiarizing selected audiences with the realities of resource development in the western United States. Until 1996, the primary activities of EMFI were two summer field programs that accepted participants from federal and state government, the media, and the K-12 community to mineral extraction sites, processing facilities, impacted communities, and related resource developments. These field programs addressed geological, technical, economic, environmental, social, institutional, and political aspects of the various issues stemming from resource development. EMFI also conducted an annual Washington Workshop and published a newsletter titled "Field Notes."

Both the public and private sectors funded these flagship programs. Each sponsoring agency was entitled to send one or more representatives to field programs. Although reduced funding from various federal agencies forced tdiscontinuation in 1997, the institute has continued to offer specialized field programs for a variety of other Mines programs. SPACE is seeking to resume the EMFI field programs and has solicited the counsel of alumni, governmental decision makers, potential funding agencies, and the natural resource industries More than eight hundred individuals are alumni of the field institutes, many of whom are still active. A positive sign was the EMFI Field Program conducted in August 2003. It focused on hydrogen fuel cell development, petroleum development, enhanced oil recovery, natural gas development, renewable energy sites, coal bed and tight gas sands development, coal-fired power plants, synthetic fuel resources, and hydroelectric generating stations.

The Mines band in formation. Always a group to behold.

purposely or inadvertently adopted an elitist demeanor toward one another or toward non-Greek students. Referred to at Mines and many campuses as "barbs" (short for barbarians), the non-Greeks often organized their own athletic and social activities. At Mines, they adopted the formal title of Barbs and have become very active in campus activities. Greek organizations, naturally, participate in a wide spectrum of campus affairs, often to the point of indispensability.

Following World War II, fraternities and sororities, like many other collegiate groups, adjusted a number of their traditional procedures to satisfy their national offices and new generations of students with new expectations for collegiate life. Fraternities at Mines in 1974 were the following:

Alpha Tau Omega (Epsilon Alpha Chapter). Founded at the University of Richmond in 1865. Established at Mines in 1929.

Beta Theta Pi (Beta Phi Chapter). Founded at Miami University of Ohio in 1839, the first fraternity west of the Alleghenies. Established at Mines in 1908.

Kappa Sigma (Gamma Gamma Chapter). Founded at the University of Virginia in 1869. Established at Mines in 1904.

Sigma Alpha Epsilon (Colorado Lambda Chapter). Founded at the University of Alabama in 1856. Established at Mines in 1903.

Sigma Nu (Gamma Eta Chapter). Founded at the Virginia Military Institute in 1869. Established in 1901, the first fraternity at Mines.

Sigma Phi Epsilon (Colorado Delta Chapter). Founded at Richmond College in 1901. Established at Mines in 1923.

On 4 November 1974, nine female students—Lori Stute, Collette Price, Claudia Roberts, Patricia Smith, Harriet Van Lieu, Clarissa George, Suzann Gonzales, Brenda Steele, and Joni Hilltrop—founded Alpha Delta, the first sorority at Mines. On 20 January 1980, Alpha Delta officially became a chapter of Sigma Kappa, a national sorority founded in 1874 at Colby College. Later national sororities coming to Mines were Pi Beta Phi in 1986 (founded at Monmouth College in Illinois in 1867) and Alpha Phi in 2000 (founded at Syracuse University in 1872).

Chapter 8: Mines at "100"

A circus of Mines students—clowning around.

Dr. Frank Hadsell (geophysicist) bares his background.

A feminine touch to the traditional tug-of-war.

This is what it is all about—Graduation!

183

CHAPTER 9

Prospects

There is nothing permanent except change.
—Heraclitus

Transformation—evolutionary or revolutionary, welcome or unwelcome, understood or misunderstood, feared or trusted, recognized or ignored—remains the basic force that shapes history. Even the caveman must have grunted or grimaced when confronted with change. For the decade leading to the millennium, the most significant political change was the end of the cold war. The fall of communism was abrupt and, with the possible exception of Mikhail Gorbachev's foresight, virtually unanticipated. New political and economic horizons for the world beckoned. Higher education, too, had its beckoning horizons: increases in enrollments, tuition, faculty salaries, endowments, and government and private research grants; but it was forced to negotiate prior economic pitfalls in the later 1980s.

I

When asked how the search for a replacement for McBride was progressing, Trustee Fred Schwartzberg quickly responded, "We are not replacing Guy McBride; we are looking for a new president." A search committee of two trustees and representatives from all sectors of Mines required eight months to reduce a list of 142 candidates to 3. Of those, the unanimous choice was Dr. George S. Ansell, an energetic administrator capable, in the committee's view, of leading Mines through necessary adjustment. In May 1984, Ansell became the thirteenth president of Mines. A metallurgical engineer, he came from his position as dean of the School of Engineering at Rensselaer Polytechnic Institute where he had received his bachelor's, master's, and doctoral degrees and where, with the exception of a stint as an officer in the U.S. Navy, he had spent his previous professional career. Hardly a caretaker administrator, Ansell left a distinct mark of success at RPI. During his tenure as dean, the School of Engineering significantly developed its graduate program, tripled the enrollment, quadrupled funded research, and became one of the ten largest engineering graduate programs in the country. He accomplished a 35 percent increase in faculty size, construction of a $15 million Engineering Center, the addition of thirty-one undergraduate laboratories, and the start of construction on a $20 million integrated electronics center.

Opposite Page: In a déjà vu photo, Nancy (Easley) Ise joins President Bickert and a group of Mines women on 3 October 1998 in the Florence Caldwell Centennial Celebration on the steps of Guggenheim. (DAVID CORNWELL)

Dr. George S. Ansell, thirteenth president of Mines.

If Bishop Randall came west from Boston to found educational institutions, Ansell came west from upstate New York to lead one of those institutions. Just as Randall found the West different from Boston, Ansell found Mines different from RPI. A western school established on the practical basis of extracting minerals from the earth and turning them into useful products, Mines grew from those narrow roots to become a multifaceted institution without losing its relationship to minerals or the mineral industries. In the 110 years before Ansell arrived, Mines passed through numerous efforts to define and redefine its mission and curriculum. A small, close-knit, state-supported school often underfunded, it also faced the possibility from time to time of being merged with the University of Colorado. Strong encouragement from its trustees, alumni, and influential supporters, however, protected the school and its mission. RPI was an eastern school, a private school with a broader mission. Though not "officially" Ivy League, it was geographically close enough to feel some of the same academic winds. Whatever their differences, both Mines and RPI were proud schools with well-earned national and international reputations.

Ansell arrived at Mines during a difficult period. Preceding the expansion of the 1990s, a recession, the shrinking of the oil industry, and a decline in the college-age population all combined to reduce Mines' enrollment from 3,250 in 1983 to 2,200 in 1989. Clearly understanding that the road for Mines was not going to be an easy one, Ansell stated in his inaugural address that the mineral and energy industries were threatened as never before:

> Of one thing we are certain: whatever the future will be, it will be different from what it is today and it certainly will not be what it was yesterday. However, the future will be determined by what we do now and what we accomplish tomorrow. We have the obligation and the capability to help create that future: for this school, for the resource industries, for this state, and for the nation.

Calling for renewal, Ansell made clear that while the school would retain its primary focus upon the exploration for, and exploitation of, the earth's mineral and energy resources, "the way in which we address this focus, the science, technology, economics and management of these enterprises is, by necessity, going to undergo major change." Past Mines presidents had called for change and more change, but perhaps not with Ansell's specificity. The agenda he outlined in his address focused on five items:

> First, we must re-examine our curriculum: what we teach and how we teach.

E-Day—band trucking along.

Chapter 9: Prospects

E-Day—the tricycle race.

Second, we must expand our research base, both to provide the base of scholarship, which traditionally supports first-rate teaching, and to develop the new technologies so desperately needed by the mineral industries.

Third, we must broaden our focus to educate our students for leadership.

Fourth, we must start thinking, educating, and operating in more of an international context.

Finally, we must educate our students for an expanded role within industry, government, and academe—one that does not begin with exploration and end with processing, but one that continues further with career possibilities that extend to the finished product.

The goal, he emphasized, was "not to lose the base of our curriculum, but to widen it; not to lose our concern for undergraduate teaching, but to strengthen it through scholarship and revitalized faculty careers" ("Benchmarks," 20 February 1985).

In the 1988–89 academic year, with key support from Ansell and strong faculty interest, Mines passed a milestone of change in its faculty-administration relationship through the formation of the Faculty Senate. Led by Dr. Dendy Sloan, professor of chemical engineering and petroleum refining, as its first president, the senate consisted of thirteen elected faculty members: eight full professors with at least five years of academic faculty experience and five academic faculty members of any rank with at least two years of academic faculty experience. At the same time, a faculty committee under the direction of Vice President of Academic Affairs Dr. John Golden produced a new edition of the Faculty Handbook that established a de jure tenure system, bringing Mines in line with most of the country's colleges and universities. Both of these events were signal changes in the relationship between faculty and administration in the governance of the school.

The senate faced its first test on 2 May 1989 when it received a petition signed by fifty-two faculty members calling for a written vote of confidence regarding administrative leadership. Strikingly simple, the ballot offered three choices: confidence, no confidence, abstain. As an agent of the faculty, the senate emphasized that it was not a participant in the process, but rather was acting according to faculty bylaws in carrying out the vote. Reported on 16 May, the results were as follows:

- 289 ballots were distributed
- 111 voted no confidence
- 30 voted confidence
- 22 abstained
- 126 did not vote.

The unknown elements in the voting were the 22 abstentions and the 126 who did not return their

These two can really cut it.

Music at Mines

That Mines has had a long and fruitful musical tradition to accompany its academic and research efforts in science and technology may surprise the casual observer. The highly regarded musical groups of today, however, have their roots in the early years of the school and are an important part of today's campus scene.

The first Glee Club concert occurred on March 22, 1912, before a large and appreciative audience in Guggenheim Hall. Directed by R. S. Hawley, the twenty members of the Glee Club presented a varied program of popular and classical songs. The hit of the evening was the "Faculty Song," a musical spoof of some members of the Mines faculty.

The Mines band came to life in the fall of 1920 with Professor A. E. Bellis, head of the Physics Department, as director. Bellis received no pay for his extra duty with the band, other than the enjoyment of developing a first-rate musical organization. Under his guidance, the band soon became well known and respected on the campus and in the community. Attired in natty military-style uniforms, the band played at all rallies and football games, gave Sunday afternoon concerts, and marched in parades. To support the band, each student paid one dollar a semester.

Music on the Mines campus has come a long way since those early days. The Glee Club has become the chorus, complete with sopranos; and the band uniforms have changed to hard hats, plaid shirts, and blue jeans. In the late 1980s, at the urging of the Humanities and Social Sciences Department, the Financial Aid office arranged for eight scholarships in music, five for the band and three for the chorus. This number increased rapidly until over one hundred students held music scholarships. Courses in music appreciation and music theory were introduced in the 1980s, and the McBride Honors Program Flute Choir made its debut in 1990.

Pictured above: The McBride Honors Program Flute Choir.

---

ballots. Forgetfulness, disinterest, absence from campus, lack of understanding of the problem—all may have been reflected in those numbers. Although a majority of the total faculty, for whatever reasons, did not register a lack of confidence in the administration, those who did indicated Mines was going through another period of faculty unrest.

Common enough in academe, votes of no confidence can cause significant complications in the operation of an institution. Fortunately, most of the Mines community preferred not to drag the school through a conflict that might take years to heal. In a press release announcing the poll results, the senate emphasized the faculty's extreme loyalty to the institution and its general acceptance of the new directions the school was taking in response to changes in higher education and the minerals industry (see the *Transcript* 25 May 1989).

Because of the brevity of the confidence ballot, the senate interviewed as many faculty members as possible to discern the motives behind the recent vote. The result was a Bill of Particulars (fall 1989) summarizing the interviews but circumspectly omitting anecdotal, individual information. Criticisms ranged among style of leadership, administrative organization, lack of communication, low salaries, and increased emphasis on graduate programs and research. Some of these issues were causes of concern and, at times, of friction, well before Ansell arrived. One section in the bill serves as a capsule summary:

> Traditionally, CSM has been primarily a teaching institution, operating under an authoritarian administration. The changes to great "democratization" of the school and increased emphasis on research came slowly until the arrival of President Ansell when both aspects were stressed as important parts of the school's new posture. Both were enthusiastically accepted by the majority of the faculty. At the same time, the proposed changes raised the expectations of the faculty. Although the President did indeed provide the vision and stimulus that the school needed, the faculty's expectations have not been realized, and the initial leadership has failed to inspire the support necessary on the part of the faculty to achieve objectives.

Ansell moved rapidly to address the issues behind the vote, working to improve communication across the campus and planning significant administrative changes. In addition to regular meetings with the

E-Day—Is this a man's work?

E-Day—Yes, it is the ore cart pull, though the cart seems to be out of sight.

senate, he held individual conversations with department heads, faculty members (over half before fall), and administrators. He also discussed the situation with the Board of Trustees and obtained its support for changes to improve both function and communication in administrative affairs. Probably most significant, department heads would henceforth report directly to the Vice President of Academic Affairs. While adversity may not always prove positive, the confidence vote in question provided an example of Mines' ability to move beyond conflict to cooperation: the senate and Ansell kept the lines of communication open; the Board of Trustees supported Ansell, convinced that he would provide the necessary leadership; and the faculty, as ever, continued to conduct its pedagogical and research responsibilities. This cooperation between administration and faculty, moreover, made it possible for Mines to enjoy a welcome period of growth in enrollment (undergraduate and graduate), faculty salaries, campus facilities, and research leadership.

Rocky Mountains to the World: A History of the Colorado School of Mines

Women's volleyball team.

Another important event requiring cooperation between the senate and the administration was Mines' removal from the AAUP censure list. Although the dismissal of faculty member Winton Laubach in 1971 that resulted in the Mines censure had essentially been forgotten, the presence of a faculty senate and the new de jure tenure policy encouraged Mines to revisit the matter. Dr. James Brown, senate president in 1991, exchanged correspondence with the AAUP regarding Mines' new tenure policy and with Laubach, who graciously said he was no longer interested in redress from Mines for his dismissal and was pleased that Mines now possessed a de jure tenure policy. On the basis of this new information, the AAUP voted at its November 1991 meeting to remove Mines from the censure list, effective 1992.

II

In the decade from 1984 to 1994, Mines sought to accomplish the goals Ansell laid out in his inaugural address. Carrying on from the McBride era, the school increasingly emphasized the concept of "Stewardship of the Earth." The Environmental Science and Engineering Division, established in 1981, maintained a limited profile until about 1986. By 1992, it was the school's largest and fastest growing graduate department with 215 students. An undergraduate minor focused on groundwater resources, soil and water contamination, mining wastes, and hazardous substances. Also, extensive research was being conducted through several centers and institutes, one of them the International Ground Water Modeling Center. In late 1992, Mines was approved to offer the doctorate in Environmental Science and Engineering. Through the remainder of the 1990s, this division grew rapidly. In 1999, it granted a total of thirty-three master's degrees and two doctorates, introduced an undergraduate option in conjunction with the Engineering Department, and developed an emphasis on applied environmental microbiology. Interdisciplinary in its thrust, the division became an integral part of the Mines curriculum.

Graduate enrollment across the campus increased from 611 (22 percent of total enrollment) in 1984 to 848 (30 percent of total enrollment) in 1994. Two-thirds were working for the master's degree and one-third for the doctoral degree. Sponsored research grew 9.3 percent to more than $14 million, almost half funded by private industry. Many areas of research were interdisciplinary in focus and crossed sections of the engineering and science departments, including the International Ground Water Modeling Center, the Advanced Steel Processing and Products Research Center, the Center for Welding, the Colorado Institute for Fuels and High Altitude Engine Research, the Robotics Automated Welding Laboratory, the Colorado Center for Advanced Ceramics, the Center of Geoscience Computing, the Center for Wave Phenomena, and the Reservoir Characterization Project. Together, these centers and institutes, along with more to follow, enhanced Mines' reputation as an institution on the cutting edge of engineering and scientific research. For a series of excellent histories of Mines' academic

The indoor track team.

departments and programs, see the 1996 and 1997 issues of *Mines Quarterly Review*. For earlier histories, see MM February 1994.

During the decade in question and continuing to the end of the 1990s, each department on the campus was constantly inching the bar higher in curricular and research efforts to prepare for a dash into the twenty-first century. As of 1998, departmental enrollments were as follows:

Chemical Engineering and Petroleum Refining
    415 undergraduate   60 graduate

Chemistry and Geochemistry
    50 undergraduate   37 graduate

Economics and Business
    42 undergraduate   96 graduate

Engineering
    762 undergraduate   63 graduate

Environmental Science and Engineering
                      85 graduate

Geology and Geological Engineering
    114 undergraduate   135 graduate

Geophysics
    54 undergraduate   55 graduate

Mathematics and Computer Sciences
    240 undergraduate   37 graduate

Metallurgical and Materials Engineering
    100 undergraduate   110 graduate

Mining Engineering
    68 undergraduate   39 graduate

Petroleum Engineering
    125 undergraduate   59 graduate

Physics
    100 undergraduate   28 graduate

Although it is difficult to establish an optimal standard of balance between undergraduate and graduate enrollments, Mines, near the end of the century, appeared to maintain a respectable

It's the Mines wrestling team of 1991 at the Edgar Mine entrance.

distribution of students among its engineering and science departments and among undergraduate and graduate enrollments. With the new undergraduate option in Economics and Business and the broad areas of study available in the engineering and science departments, many of them interdisciplinary, Mines was offering, for its size, a wide choice of options for entering students.

Two new programs having broad implications for the campus were gaining momentum when Ansell arrived. The first, a spin-off from a $350,000 National Endowment for the Humanities grant awarded to the Humanities and Social Sciences Department was an honors program begun in 1978. With Dr. Tom Philipose's vision and guidance as the first principal tutor, a post he held until the spring of 1989, the program, funded by an additional endowment of $100,000 in honor of former president McBride, quickly and effectively began to make its mark in the Mines curriculum as the Guy T. McBride Jr. Honors Program in Public Affairs for Engineers. Open by selection to second-semester freshmen, it required twenty-four semester hours that were team-taught by faculty from sciences and engineering and from the humanities and social

Professor Dale Foreman of the Engineering Department instructs students from Denver high schools in the principles of surveying in a special summer course.

sciences. Approximately thirty-five to forty students were selected each year. During the summer of their junior year, students exercised the option of an internship in government or industry or six weeks of foreign study. Later, a public service summer option was added. Students completing the program received a certificate and a minor. In 1991, the program competed for, and received, a CCHE Excellence Award of $450,000 distributed over three years. Overseen by the principal tutor and the Tutorial Committee of faculty members from across the school, the program achieved national notice for blending the humanities and social sciences into the engineering and scientific thrusts of a Mines education. McBride has continued to support the program monetarily. The program stands as a fitting tribute to McBride's vision of engineering education specifically and collegiate education generally.

The second program, Engineering Practices, Introductory Course Sequence (EPICS), was a two-semester set of courses for freshmen and sophomores. Like the Honors Program, it was also team-taught by faculty from engineering and science and from the humanities and social sciences. Under the initial direction of Dr. Richard Culver, assisted by members of the Military Science Department, the course began as a pilot project in the early 1980s. Later, under Dr. Michael Pavelich's leadership, it became a permanent requirement for all undergraduates. Utilizing an open-ended problem-solving team approach to projects sponsored by off-campus clients from industry and government, EPICS left its stamp on a broad range of Colorado agencies and industries. In addition to giving students an early immersion in design, the course provided ample practice in written and oral communication and computer usage. Still a vital part of the curriculum today, EPICS has produced students with the capability to study and solve engineering problems and to communicate their oral or written results clearly and effectively. The skills learned by the students are invaluable in their other courses and in their future professional careers.

The Humanities and Social Sciences Department was not only active in Honors and EPICS, it was achieving recognition in its own right. In 1983–84, the department received a $300,000 endowment from the Atlantic Richfield Oil Company and, in 1990, a $270,000 highly competitive Challenge Grant from the National Endowment for the Humanities, the latter to be matched three-to-one by Mines. Most schools receiving such grants take from three to four years to meet the match. Mines, however, met its match in nine months with gifts exceeding $810,000 from Ralph Hennebach '41 ($317,000); Ahmed Kafadar ($360,000); and ASARCO Inc. of which Hennebach was CEO

Three Mines presidents—McBride, Ansell, and Childs—reflect decades of service to Mines.

Chapter 9: Prospects

The Careers of Ted P. Stockmar

Ted P. Stockmar '43 has had two successful careers, one in law and one in service to the Colorado School of Mines. As an undergraduate, he had two interests—petroleum engineering and football—and he excelled in both.

Earning a law degree from the University of Denver in 1948, Stockmar joined the law firm of Holme Roberts and Owen and served as a partner for forty-one years. Also in 1948, still in his twenties, he became a member of the Mines Board of Trustees, serving in that capacity for thirty-five years, including ten years as president from 1969 to 1979. Nothing important it seems was short-term for Stockmar. His present position is of Counsel with Holme Roberts and Owen.

As a Mines trustee, Stockmar was an active proponent of the Humanities and Social Sciences program and was active on the visiting committee of the Guy T. McBride Honors Program in Public Affairs for Engineers. He also spent thirty-five years on the Colorado School of Mines Foundation (two terms as president). Author of many articles relating to natural resources law and economics, he served ten years as a member of Governor John Love's Oil Shale Advisory Committee and was formerly Colorado's representative to the legal committee of the Interstate Oil Compact Commission.

Mathematics was one of Stockmar's favorite subjects, and he often engaged people in conversation on that topic. Invariably, he asked them if they knew the numbers for pi. Of course, most had no idea. From memory, Stockmar would enlighten them with 3.1415926535. Stockmar also had some of the old Mines spirit, as evidenced in an incident following World War II when he buzzed the Mines campus in a four-engine prop bomber, vibrating buildings and cracking windows—a deed that he did not admit to for many years.

In 1971, Stockmar was named an honorary member of the Mines Alumni Association. In 1987, he received the Mines Distinguished Achievement Medal and in 1989, the Mines Medal. In 1997, Mines presented Stockmar with its highest award: an honorary doctoral degree.

($250,000). This grant and the matching gifts made it possible to establish the Hennebach Visiting Professorship, to sponsor the Artist-Scholar lecture series, to increase library holdings dramatically, and to provide funds for faculty development. These sums, while not of the magnitude of many of those generated by the Engineering and Science Departments, brought the department's endowment to over $1.5 million, an amount large enough to dispel any notion that Mines was not supporting the humanities and social sciences. This endowment total has grown even more in the last few years through the continued generosity of Hennebach.

In 1996, the Global Systems and Cultures Department and the Latin American Center were established to enhance and increase international efforts at Mines. In addition to varied consulting activities, the department successfully developed a curriculum with a strong focus on the political economy of areas in Latin American and Southeast

Rocky Mountains to the World: A History of the Colorado School of Mines

Arthur Lakes Library framed by evergreens.

Asia, providing a minor and a certificate for interested students. This department and the Humanities and Social Sciences Department combined in 1991 to become the Liberal Arts and International Studies Division, which today is breaking new curricular frontiers in providing an effective program in the humanities and social sciences for Mines students.

III

And what of entering freshmen during the period in question? Mines' average ACT and SAT scores were 27.3 and 1234, respectively, in 1997. Colorado State's averages were 24.2 and 1114 while the University of Colorado at Boulder's were 25 and 1157. Average high school GPA of freshmen at Mines was 3.8; at Colorado State, 3.5; and at the University of Colorado at Boulder, 3.4. Percentage-wise, Colorado State admitted the highest number of in-state freshmen with 72 percent; Mines was second with 66 percent; and University of Colorado at Boulder was third with 57 percent. The average numbers of years to complete a bachelor's degree were as follows: Mines, 4.9; Colorado State, 5.1; and University of Colorado at Boulder, 4.8. Probably because its tuition was the highest, Mines was granting a higher percentage of financial aid to entering freshmen than either of the other schools. (See "How Mines Stacks Up," the *Oredigger* 27 April 1999.)

Once students arrive on campus, no matter how able they are, retention is of primary concern to every college or university. Losing students is very much like losing blood—with transfusions not so easy. Because Mines' retention rate in 1997 was a somewhat uncomfortable 60 percent, the first study of the campus culture was sponsored that year by the CSM Retention and Management Committee and carried out by a subcommittee on academic culture. A ninety-eight-question survey was completed by 826 students representing all classes. Dr. James H. Banning of Campus Ecology Services in Fort Collins conducted a second study in 2002. Though not scientific studies, they were quite similar in their conclusions. Both, while acknowledging many strengths in Mines' culture, noted several significant problems, as indicated by the following from the 1997 report:

> But despite these positive indications, rumblings of discontent are emerging among the student body. Some consequences are that affected students leave, or they develop sullen attitudes which do not bode well for our

future relationships with alumni or for our attractiveness as a choice institution. This subtle phenomenon resides in our collective yet nebulous academic culture. It resides in the relationships which we have with students whose expectations are increasingly consumer-oriented. It resides in the contrasts between partnered learning among adults and subservient instruction to kids. It resides in the attitudinal differences between blatant arrogance of acquired stature in a profession and the mentored guidance of new people into that profession. It has everything to do with our interaction with students, and it concerns both personal interaction and operational interaction with academic systems. Two-way expectations in the student-academy relationship are the prerequisite for a healthy academic culture.

Stained glass window in Alderson Hall.

The 2002 report referred to the above passage, noting that the campus atmosphere had not changed significantly from that of 1997. Some students viewed the campus culture as "unhealthy," "screwed up," "callous and cold," "boring," and "stressful." A sampling, moreover, viewed the faculty as "wanting to put you through hell," as "gods to be worshiped but can't teach," as "dictators," as "wanting to weed out people," as "insensitive," as "wanting to kill you with homework," and as "out of touch with the real world." Among concerns outlined in the report were problems of students staying on track to complete their programs, lack of organization in the scheduling of classes, common-examination hours interfering with schedule flexibility, and too much departmental power in scheduling classes. The 1997 review reported the following:

> Our academic culture is expressed through some of the deepest traditions of the institution. These traditions combine within a suite of educational programs which are producing competent and successful graduates. While this is gratifying, evolving expectations in society and in our student clientele are causing some of our modes of behavior to be outdated and frustrating to students. The mechanics of how courses are

The entrance to a renovated Alderson Hall.

> ## The Senior Stetson
>
> One of Mines' oldest traditions and one that represents the expression of pride in a profession is that of the Senior Stetson. Dan Witkowsky '66, in MM of September/October 1995, covers in detail the interesting story of the Stetson with its unique block and narrow brim, setting it off from the popular cowboy hat.
>
> According to Witkowsky, mining engineers during the gold and silver booms of the early West wore silver bellow Stetsons as a unique symbol of their proud profession. It was a visible badge of distinction. Mining engineers graduating from Mines and other mining schools were easily identified by the Stetsons they wore. Jesse Morgan in his book noted that the Mines band in 1921 adopted the Stetson as part of their uniform. From that time until the 1970s, the Stetson, and only the Stetson, was a vital part of Mines lore. No other hat would do. Witkowsky felt that wearing "a box in which a Stetson came was more comfortable than wearing a non-Stetson hat."
>
> Cottrell's in downtown Denver was the primary source for the purchase of Stetsons by Mines seniors. A Cottrell's Stetson was further enhanced with "The Senior, Colorado School of Mines" stamped in the lining. Witkowsky noted the ubiquity of the Stetson:
>
>> One Senior Stetson was left at the Vietnam Memorial in Washington, D.C., in tribute to Ron Layton '66, the only CSM graduate known to have died in the Vietnam conflict. The Stetson was autographed by all those present at the twenty-fifth reunion of the class of 1966 in 1991, and on the underside of the Stetson brim was later written. "To Ron Layton, The Colorado School of Mines will never forget your sacrifice. You're a helluva engineer—the Class of 1966. May 10, 1991."
>
> More than a year later, on the evening of October 8, 1992, the Stetson was left below Panel 27W of the Vietnam Veterans Memorial. By leaving the Stetson, the class of 1966 ensured that their classmate would never be forgotten for his effort.
>
> Although the tradition of the Stetson, despite alumni efforts to the contrary, may have faded as years passed, it has earned its niche in the history of Mines.

scheduled and the subtleties of how we teach our courses are viewed as the most troublesome issues.

While Mines has always been proud of its academic rigor, the above criticisms are not unique to it. Other studies have shown that students and alumni from a wide spectrum of colleges, universities, and professional schools hold the same views. The 1997 report, for example, referred to a national study by Elaine Seymour and Nancy Hewett, *Talking About Leaving: Why Undergraduates Leave the Sciences*, that shows 93 percent of students leaving engineering do so because of poor teaching. The 2002 report stresses that an improvement of the cultural climate at Mines is the responsibility of Mines. Taken together, these reports candidly presented urgent concerns and a significant number of remedial steps to improve campus culture. Both emphasized the importance of strong leadership and effective feedback from all sectors of the Mines community. While these may not seem giant steps, they must be continuous and coherent in their direction. Among Banning's recommendations were the following:

- Addressing the possibility that some academic stress may be unnecessary and is being promoted out of misguided efforts of instructors.
- Reducing the number of hours required to get a degree.
- Giving students more time off.
- Increasing co-curricular, on-campus activities.
- Establishing an academic ombudsman's office.

Mines' relatively small size should help in bringing about more attention to, and reduction of, the problems of stress, whatever their sources. To Mines' credit, it has never been a school to obscure problems. On the contrary, it has followed the philosophy that only by recognizing problems can solutions be reached.

IV

During the 1980s, the number of female students was growing rapidly, with the average graduating class having 91 women. *A Century of Women at Mines* mentions two basic causes for the increase: (1) the women's movement in America and (2) the changing curriculum to include mineral economics, environmental science, and basic engineering programs. The number of women graduates rose to 126 in 1986 and then declined along with the male enrollment to 74 in 1990. By 1997, however, women made up 25 percent of the enrollment and a significant part of the Mines community. *A Century of Women at Mines* records a number of women's reactions as they strove to succeed in a school that had been a male bastion since its origin. One of the best is about Dr. Ramona Graves, the first woman in the United States to obtain a Ph.D. in Petroleum Engineering and now a professor at Mines: "I get real tired of 'look what women have done' stories. I never think about my gender. I've often been the only woman in a class or at a meeting, but I never think of myself in those terms." The women at Mines today undoubtedly agree. Indeed, women have more than held their own in classroom grades

The renovated Ben Parker Student Union.

The "M" climb with rocks in hand (or arms).

and commencement awards. They participate in varsity sports, the band, the chorus, Mines Little Theater, the Board of Trustees, the *Oredigger*, and many more activities. *A Century of Women* phrases it as follows:

> Female students at the Colorado School of Mines have finally come into their own. The number of women on campus continues to grow and the number of women in the science, engineering, and math fields continues to grow. Although we are still far from a 50/50 split between women and men, the number of women in these fields continues to grow. In the future, the Colorado School of Mines may be able to boast having 50 percent of the students on campus women.

Two organizations of considerable importance to women at Mines are Women in Science, Engineering, and Mathematics (WISEM) and the Society of Women Engineers (SWE). They assist first-year women in their acclimation to a predominantly male school with a mainly male-oriented ambiance dominating the campus, and they provide continuing support to upperclass women throughout their tenure at Mines. Polls have shown that most women at Mines have found meeting the challenges of the school to be worth the effort and even enjoyable. As one first-year student noted, "There is a lot more to growing up and becoming an adult than seeing another candle lit on your birthday cake."

As Mines was gaining in numbers of women students during the 1990s, it lost a traditional and honored component of its curriculum. On 12 July 1990, Ansell received word that because of a reduction of the armed forces and the accompanying decline in the need for officers, the ROTC unit at Mines was to be deactivated at the end of the 1990–91 school year. The University of Colorado absorbed the programs of Regis University, University of Denver, Metro State College, and Mines. Although the political and moral support Mines mounted did not prevent the closure of its storied program, it enabled the school, under the aegis of the University of Colorado, to maintain its own Military Science Department, in which both Army and Air Force cadets can work toward commissions. The echoes of cadence-counting and the color of the precision parades on Kafadar Commons may have faded, but what the 115th Regiment of Army Engineers accomplished in seventy-two years of war and peace is indelibly recorded in the annals of American military history. In that sense, and through the concerted efforts of the present Honorary Colonels of the regiment, the 115th will never be deactivated.

It's not exactly where you splash the whitewash but how much you splash.

The "M" is painted and so are the painters.

Because of the many Mines graduates who sustain vital roles in the mineral and energy industries of the world, Ansell sought to increase the global perspective of students still on campus. In addition to the many individuals and groups coming to Mines, the 1990s brought increased formal exchange agreements with a number of foreign universities to provide student and faculty opportunities for study and research abroad and to bring foreign students and faculty to Mines. Included are the University of Guanajuato, Mexico; Lulea University in Sweden; the Mining University of Leoben, Austria; Istanbul Technical University, Turkey; and the Monterrey Institute of Technology, Mexico. Given its reputation and past experience and the demands for trained engineers and scientists on the global scene, Mines remains committed to sending graduates from the Rocky Mountains to the world.

The decade also exhibited a pragmatic side. Ansell's efforts to upgrade the physical plant strengthened Mines in both beauty and function. Rather than appealing to the legislature for new buildings, not an easy task, he followed the strategy, at least in the beginning, of seeking funds to improve present structures. Virtually all the main buildings were expanded and/or renovated at a cost of $52,125,911. Among the major buildings profiting from his attention (with completion dates) were Stratton Hall (1986); Berthoud Hall (1988); Steinhauer Field House (1990); Engineering Hall (1991); Green Center (1992); Alderson Hall (1994); Coolbaugh Hall (1995); and still-to-be-funded Hill Hall (2000). Additional projects in the planning stages during Ansell's final years were a computer-oriented classroom building, Mines Park (student housing), and renovation of the Parker Student Center. Bricks and mortar translated significantly into improved teaching and research.

During his tenure at Mines, Ansell, like McBride, proved a successful fund-raiser in both the public and private arenas. At a 16 November 1991 gala celebration at the Brown Palace Hotel in Denver, hosted by Ansell and his wife Marjorie; National Campaign Chairman and Homestake Mining Company Chairman and CEO Harry Conger and his wife Rosemary; Adolph Coors Company Chairman Bill Coors and his wife Phyllis; and Texas Instrument Company founder Cecil Green, Mines

A sport for all seasons.

announced RESOURCES, a capital campaign to aid the school in carrying out its mission for the present and for the future. The goal of the five-year campaign was $60 million. Prior to the public announcement, approximately $17 million had already been raised. Projected use of the funds encompassed student and faculty support, equipment, library, annual support, and endowment. After five years, RESOURCES had surpassed its goal by $13 million. David L. Powers, vice president for Institutional Advancement, noted in a memorandum to the Colorado School of Mines Foundation Board of Directors, "The achievement of the RESOURCES Campaign, $72,784,990 against a goal of $60,000,000, speaks for itself as an expression of loyalty, encouragement, and a belief in the School's leadership and future."

Addressing a meeting of the Denver West alumni section on 12 October 1994, Ansell reminded his audience of his earlier remarks regarding changes necessary in the Mines mission and curriculum and forecast how he saw them evolving. "The biggest change sweeping the Colorado School of Mines—and other colleges like CSM," he said, "is the shrinking demand for specialists in applied natural resources fields and the impact that it is having on how those specialists are educated and trained." All Mines students no longer have the former strong orientation to be metallurgists or mining engineers, as evidenced by the rise of engineering per se to be the largest undergraduate program at the school.

Granted, Mines might be the largest or second largest American school in almost all of the resource discipline programs at both the undergraduate and graduate levels. "What's different," he noted, "is that being the largest in relative terms is no longer being the largest in absolute terms" (MM November/December 1994). Although geophysics, metallurgy, mining, and petroleum might not return to their former enrollments, they would certainly remain steady and healthy. Ansell predicted that disciplines in the natural resources would become primarily graduate programs, enhanced by students arriving from diverse undergraduate backgrounds. He saw two things emerging from the curriculum study underway: (1) a broadening of the undergraduate curriculum toward a more common, core curriculum; and (2) a realization that the education for a bachelor's degree is only a part of the education sequence. (For more on the curriculum study, see "Remodeling the Curricula" in MM November/December 1996.)

During Ansell's last year as president, Hill Hall had been undergoing a major renovation, that, when finished, placed Mines at the top in metallurgical facilities among engineering schools. With its

Baby, it's cold outside!

Mines invaded by a dragon during the Asian Festival.

completion in the fall of 1998, Metallurgy became the George S. Ansell Department of Metallurgical Engineering and Materials Science. In addition, the George S. Ansell Distinguished Chair in Metallurgical Engineering was established. More than $8.2 million was raised in a special campaign to fund this chair and associated educational programs. Frank Erisman, president of the Mines Board of Trustees, summarized Ansell's achievements: "Dr. Ansell's remarkable service to Mines has included numerous accomplishments over his fourteen-year tenure. He will be leaving a legacy of dynamic leadership and strength" (MM September/October 1998).

It is worth repeating that while presidents of colleges and universities take credit and/or blame for whatever changes or lack thereof their institutions undergo, ultimately the faculty provide the educational continuity that is vital to any school. For every Nobel Prize winner, thousands of faculty members conduct their teaching and research out of the limelight. The professorial hoods symbolize endless hours of purpose, preparation, practice, and postmortem. Socrates had no administrators, no fund-raisers, no trustees, and no staff to support his efforts. Nor did long-range plans or tenure concern him. He truly was Ralph Waldo Emerson's "man thinking," a mind sharpener, whose knowledge and the desire to share it cost him his life. But he paid without complaint. Although today's educational reality, unlike that of ancient Greece, demands a complex, bureaucratic support system for a school to accomplish its mission, the foundation of stability and growth remains the faculty, individually and collectively.

V

As it was one hundred years earlier, Mines was ready and eager to move into a new century. Change in the past twenty-eight years had brought the school through some trying times to positions of strength in endowment, funded research, physical plant, total enrollment, faculty quality, and international reputation. All academic departments, moreover, were confident in their long-range plans and in their abilities to institute them. To lead the school into the new millennium, Mines sought for its new president a proven educator with a broad and successful background of teaching and administration. The man chosen was Dr. Theodore Bickert.

Bickert's undergraduate and graduate degrees were awarded by the highly respected Johns Hopkins

Asian music at the Asian Festival? Yes.

They all came to the pig roast on E-Day.

University. Before coming to Mines, he had achieved a long and fruitful career in education, research, and administration at the collegiate level. After two years in military service, assigned to the Satellite Communications Agency at Fort Monmouth, New Jersey, Bickert transferred to Syracuse University, where he spent twenty-one years as a professor and five years as Dean of Engineering. As an electrical engineer, he focused his research interests on developing and applying analytic techniques for studying stability and performance of electrical networks and control systems. He taught a broad spectrum of courses in electrical networks and control systems, including digital circuits and systems. From Syracuse, Bickert went to Michigan State University in 1989 as Dean of Engineering where he remained until coming to Mines. Fond of trekking and mountain hiking, the "freshman" president participated in one of Mines' oldest traditions, carrying a ten-pound rock up to the "M" with his "fellow freshmen."

In two addresses to alumni groups, Bickert spoke of his goals for Mines. In one, he recalled being told by a former colleague at Syracuse that he would have a world-class group of geologists, geophysicists, and earth resource scientists to work with. Bickert also described his vision for Mines, noting first that "our traditional disciplines will continue to be, by our collective energies, world class." Furthermore, the school will be embarking on "new endeavors, expanding our scope and positioning ourselves at the forefront of technology and innovation." Among the areas Bickert specified as central to his educational vision for Mines were the globalization of engineering, the importance of engaging students in public policy issues, and the commitment to upholding the auspicious traditions of Mines. Bickert believed that incorporating "stewardship of the earth" into the Mines mission would catapult the school onto "the high road to fulfilling an engineer's obligation to improve the human condition." He also predicted that the new Mining and Earth Systems Engineering program would enable engineers with backgrounds in geology, geophysics, and civil engineering to undertake graduate study in underground construction, excavation, and tunneling without completing an undergraduate mining degree (see MM September/October 1998).

Bickert's 1998–99 Report reflects a school successfully carrying out its mission and pursuing new areas of endeavor. In addition to capsule summaries of educational and research develop-

Three smiling Mines students enjoy an outdoor snack.

Chapter 9: Prospects

Weaver Towers

On 8 September 1993, a prominent landmark on the Mines campus received a new name. The Twin Towers residence hall became the Gaylord C. and Phyllis Weaver Towers. Gaylord "Buck" Weaver '26 was one of the school's most dedicated benefactors. Prior to his death in 1987, he arranged for $4 million from his estate to be given to Mines upon his wife's death. Mrs. Weaver died in 1990, at which time Mines received the largest endowed fund in its history. Of the endowment, $3 million was placed in an unrestricted endowment and $1 million funded the Weaver Chair in Chemical Engineering and Petroleum Refining.

Growing up on a ranch in Kremmling, Colorado, Weaver joined the United States Gypsum Company following his graduation from Mines. In 1927, he worked in Peru for the Cerro de Pasco Copper Company. Following a year as a teaching fellow at Mines, he held various positions with the Potash Company of America in Carlsbad, New Mexico, where he and his wife made their permanent home. There he founded the Weaver Drilling and Minerals Research Corporation. He also worked as an independent consulting engineer for the Duval Company, Gulf Oil Corporation, and Southwest Potash Corporation. Among his discoveries was a potash reserve near Carlsbad and three billion tons of copper ore in Arizona. In the 1960s, he became one of the largest independent oil producers in Texas. Along with his engineering successes, Weaver maintained close ties with Mines until his death at the age of eighty-five.

ments and plans for the future from each academic department on the campus, other achievements and important events are noted:

- International efforts were made to increase Mines' overseas partners and to provide study opportunities for students in Australia, Austria, China, England, France, Japan, Germany, Mexico, the Netherlands, Spain, and Sweden.

- Foreign embassies in Washington, D.C., were visited to enhance graduate recruiting in Egypt, Kuwait, Oman, Qatar, Saudi Arabia, Thailand, and the United Arab Emirates.

- Mines was named one of the top public national universities by *U.S. News & World Report*.

- The Division of Engineering received an Excellence Award of one million dollars from CCHE to enhance its curriculum efforts. Part of the award was designated for new learning techniques with the K-12 community through workshops arranged by the office of Special Programs and Continuing Education (SPACE), which maintains the largest teacher-training program in the state, enrolling 3,700 teachers.

- Mines set a record for fund-raising with more than $22 million in gifts and pledges for the fiscal year 1998. The school's endowment reached $127 million.

- A dental clinic was established on the campus, open three mornings a week.

- The new Mines Park student housing project was completed in 1998 at a cost of $5.5 million. Capacity is approximately 200.

Dr. Theodore Bickert, fourteenth president of Mines.

- A 13,000-square-foot addition to the Ben H. Parker Student Center was planned.

- Sponsored research projects totaled $20,374,902.

Accompanying Bickert in his first year was a freshman class of 565, one of Mines' best entering classes. The average SAT score was 1240, and the average ACT was 27. Sixty percent of the class graduated in the upper 10 percent of their high school classes. Total applications, according to Bill Young, director of Admissions, were at an all-time high, with Colorado students representing 70 percent of the class. Ninety minority students and 160 women evidenced Mines' efforts to increase the diversity of its student body.

During the first quarter century of Mines' history, 1874 to 1900, both the times and the school were much simpler than in the last quarter of the twentieth century. In the former period, the dominant challenge of existence was readily recognized and effectively met. The challenges facing Mines in the twenty-first century are daunting but welcome for an institution that has earned its place in higher education. Just as the early Mines enjoyed a strong external support and internal leadership, so too did the Mines of the last quarter of the twentieth century.

President Ansell presents the school medallion to President Bickert.

Twenty-fifth-year reunion of the 1974 Centennial Class.

Chapter 9: Prospects

The Mines orchestra in a practice session.

Hill Hall opens following renovation.

Action during a Mines Little Theater performance.
Guess the play.

New student housing in Mines Park.

Another academic year ends.

And "books are shutter than books can be"—
and new careers beckon.

Epilogue

It is not enough that you should understand about applied science in order that your work may increase man's blessings. Concern for man himself and his fate must always form the chief interest of all technical endeavors. Never forget this in the midst of your diagrams and equations.
—Albert Einstein

The Mines campus on a sunny Colorado fall day in 2003 reflects a simple beauty that belies the intense academic and research efforts going on in its classrooms and laboratories. Engineering Hall, Stratton Hall, Chauvenet Hall, and Guggenheim Hall connect Mines to its early years while the later buildings attest to the steady growth that has brought the school to a level of educational accomplishment and excellence that Bishop Randall could hardly have imagined. While the road has not always risen to meet Mines, and the wind has not always been at its back, the school has moved steadily toward its recognition in June 2001 by the Colorado State Legislature as an Exemplary institution (Senate Bill 01-229). The granting of this status was based on a long list of achievements recorded by Mines and the confidence of the legislature that the school would successfully carry out the terms of its Performance Agreement with CCHE. As an Exemplary institution, Mines still receives state support but enjoys greater freedom in curricular development and in setting tuition rates.

Senate Bill 01-229 also authorized Governor Bill Owens to appoint an advisory board to the Mines Board of Trustees to aid in developing and enriching the school. Those appointed were Nathan Avery '56, Chairman and CEO, the Galveston-Houston Company; D. Allan Bromley, Sterling Professor of Sciences and Dean of Engineering, Yale University; Bruce Grewcock '76, President, Chief Operating Officer, and Director, Peter Kiewit Sons, Inc.; C. L. Henry, Chairman, President, and CEO, Johns Manville Corporation; Patrick James '68, Natural Resources Management Consultant; Kristina Johnson, Dean of the Pratt School of Engineering, Duke University; Robert McKee III '68, Executive Vice President of Exploration and Production (retired), ConocoPhillips; Ralph Peterson, President and CEO, CH2M Hill Companies Ltd.; Greg Stevinson, President, Denver West Realty; and Richard Truly, Director, National Renewable Energy Laboratory.

The unique arrangement between Mines and the state established by Senate Bill 01-229 has been erroneously considered a step toward privatization of the school. Such a development would seem unlikely, given the tremendous costs involved, as well as a bit ironic, since the normal direction among colleges and universities is from private to public. Mines, moreover, is pleased with the honor and recognition granted it by the legislature and is busy identifying and sharpening its educational and research efforts in a difficult economic environment.

With Bickert's retirement in the spring of 2000, Dr. John Trefny, Vice President of Academic Affairs and Dean of Faculty, took over as Interim President during the ensuing presidential search. In August 2001, the Search Committee turned to him to serve as the fifteenth president of Mines. Trefny received his B.S. degree from Fordham University and his Ph.D. from Rutgers University. Prior to joining the Mines Physics Department in 1977, he served at Cornell and Wesleyan Universities. Trefny was head of the Physics Department at Mines from 1990 to 1995, when he was named Vice President of Academic Affairs. As with Ansell and Bickert, Trefny's background was essentially that of an academician. During his career at Mines, he has maintained a keen interest in enriching science curricula and in developing well-received programs of teacher enhancement in the public schools.

The 2002–03 academic year was one of draconian budget cuts for all Colorado institutions of higher education, as the state struggled to balance its own budget. It was not an easy time for a new president of any college or university. Determined not to let this economic crisis derail long-term goals for the school,

A Last Look at Mines Athletics

The oldest member of the Rocky Mountain Athletic Conference (as of 2002), Mines won the football championship ten times and tied for it once: 1904, 1905, 1906, 1907, 1912, 1914, 1918, 1939, 1942, 1951, 1958 (tie). The Orediggers also suffered seven winless seasons: 1919, 1920, 1932, 1934, 1957 (3 ties), 1964, and 1973. As of 2002, Mines gridders had won 340 games, lost 633, and tied 33. The records of 2001 and 2002 teams—both 7 and 4—were the best since 1951 (7 and 2).

Mines basketball teams (men) over ninety-one seasons compiled a record of 650 wins and 1,165 losses. The first season (1911–12) produced 6 wins and 1 loss, still the best percentage. Since the 1979 season, the hoopsters have posted 262 wins and 370 losses, a much better overall percentage than that of the previous seasons. The following seasons were particularly good for Mines basketball:

1979–80	16–10
1980–81	21–6
1981–82	18–7
1987–88	20–11
2002–03	18–10

In the last season above, Stephen Bahl led the Orediggers with 22.5 points per game. The low point during this later period was the 1991–92 season when Mines finished 0–27.

The women's basketball team, led by Kim Good and Marissa Eichhorn, greeted the twenty-first century with 16 wins and 11 losses, the first winning season for the women at Mines. Not to be outdone, the soccer team (men) finished the 2000 season with 14 wins, 4 losses, and 4 ties. Metro State was ranked sixth in the nation (Division II) when Mines "kicked" them 6 to 2 at Brooks Field. But the Orediggers lost in the tournament to Fort Lewis College.

Mines Moves Ahead in Funding

The fiscal year 2002 marked a high point for Mines in research funding. Awards from federal, state, and private sources totaled $30,301,850—a jump of $2.3 million from the previous year. More than 250 of the 475 awards came from the private sector, amounting to $11.6 million.

Granting $4.9 million, NASA was the largest federal contributor, followed by the National Science Foundation, with $4.1 million. The Department of Energy, the Department of Defense, and the Environmental Protection Agency also awarded grants.

Leading the Mines consortiums or centers were the Center for Commercial Applications of Combustion in Space, $4,611,470 in funding; the Reservoir Characterization Project, $1,174,731 in funding from twenty-four sponsors; and the Center for Wave Phenomena, $1,055,500 in funding from twenty-six sponsors. Given the size of Mines, these figures reflect positively on the school's efforts to maintain its reputation in both teaching and research.

Trefny assigned Dr. Nigel Middleton, Vice President of Academic Affairs, the task of preparing a report on academic planning for CCHE. In his introductory statements, Middleton cautioned that the report should be interpreted as one contributing element within a larger framework of institutional planning:

> Most notable is the umbrella activity in strategic planning, which seeks to emphasize the future of the institution in mission-related niche areas. An internal strategic planning task force has advanced this process to the point of appointing an external consultant to finalize its work and articulate the institution's broad strategic directions and accompanying programmatic and operational objectives. It is fully anticipated that these will fall under the general headings of engineering education, energy, environment, materials, natural resources, and computational science and engineering.

Middleton explained that planning would be influenced not only by the expectations of the Performance Agreement with CCHE, but also by the school's desire "to herald and demonstrate nationally the potential of a more flexible relationship between a public institution of higher education and state government." Following the enactment of the Performance Agreement, Mines initiated three new degree programs, two under the Individualized Interdisciplinary Graduate Program (Ph.D. and Master of Science) and the third, a new midcareer Professional Master of Petroleum Reservoir Systems degree. Since they were new configurations of existing courses, none of these degrees required additional faculty or courses.

Despite the specter of budgetary limitations, Trefny in an interview was both informative and optimistic.

Put your little foot, put your little foot right out.

Epilogue

In addition to emphasizing the importance of Mines' new relationship with the state, he mentioned a number of areas receiving attention:

> These focus areas have emerged from the strategic-planning process and will help define the school's growth in the future. The field of engineering education, which is our central purpose, is one focus area. Others include energy; environment; materials; computational science and engineering; and finally mining and underground construction, along with related fields. These are areas where Mines has special capabilities and a longstanding reputation. They relate to research and teaching across the entire campus. The field of energy includes fossil fuels, renewable energy technologies, economic policy and management. (MM Winter 2003)

Trefny also stressed the significance of Mines being chosen from a dozen prestigious institutions to create a university in Abu Dhabi to train a workforce for that nation's petroleum industry. "It's an educator's dream," he said, "to build a whole university from scratch with all the necessary resources provided. It doesn't happen very often. I think it's a testimony to the quality of what happens here in Golden that we were selected by a group halfway around the world." Trefny, moreover, expressed optimism regarding the Center for

Beta Theta Pi Homecoming float.

Engineering Education completed in 2000, one of only four such centers in the United States. According to Trefny, the center has attracted world attention, realizing several seven-figure grants and contracts in less than two years.

With regard to the current fund drive, "Transforming Resources: The Campaign for Mines," Trefny admitted that the title was a "play on words." More than that, he continued, the name "captures the school's mission from a more global perspective. Also, the fields we serve, the natural resource industries particularly, are evolving rapidly, and the name captures some of how Mines is changing in response to the needs of society." At a black-tie gala, attended by more than two hundred members of the Mines community, the largest fund drive in Mines history—$125 million—was formally announced. Led by cochairs Steve Chesbro '64 and Howard Janzen '76 and '77, the drive had already raised $65 million, including $2.5 million from Chesbro and his wife Dollie to establish the Chesbro Distinguished Chair in Petroleum Engineering. If past fund drives are any indication, Mines will reach, and probably surpass, the goal of the campaign.

Because of its designation as an Exemplary institution, Mines was spared some of the financial discomfort suffered by other state institutions in the past year. Still, the school experienced cuts of almost $3.5 million. Also, the lackluster performance of the

Blaster leads the parade.

Cheerleaders reflect the Mines spirit at the Homecoming football game.

stock market reduced money available from endowment funds. A letter from Trefny to the Mines community outlined budget reductions taken:

> Several administrative and service positions were eliminated and the incumbents in them laid off; other vacant positions were left unfilled. There were no salary raises for faculty and staff. We eliminated the use of credit cards for tuition payments because the service fees amounted to about $160,000 per year. The result for the current year is a lean but balanced budget, with minimal impacts upon the core educational and scholarly activities of the campus. *(Oredigger* September 2003)

Among the more positive items included in the letter were the renovation of the Maple Street dormitories, the construction of three sorority houses, the doubling of Mines Park capacity, and the only state-funded new capital-construction project for the 2003 fiscal year: a central computing facility to be built adjacent to the newly completed Center for Technology and Learning Media.

The Registrar's Report (Spring 2003) indicated a slight increase in overall enrollment from 3,077 in the spring of 2002 to 3,193 in the spring of 2003. The percentage of in-state students continued to grow, totaling 73.1 percent. The number of international students increased from 344 in the spring of 1999 to 365 in the spring of 2003, with sixty-eight countries represented. As noted earlier, any reduction in international and out-of-state students reflects negatively on the tuition generated. Minority trends ranged from a high of 12.8 percent in the spring of 1999 to 9.9 percent in the spring of 2003. At 25.7 percent, the total of women students showed little change over the past five-year period. Women represented 30.9 percent at the graduate level in the spring of 2003.

In the spring of 2003, Mines hired an independent consultant to assist in developing a strategic plan for the future. As this plan unfolded, concern arose among some faculty members and students that Mines was planning to restrict graduate research while doubling the undergraduate programs and increasing the professional-degree programs. Trefny, in a memo to the Mines community of 8 September 2003, restated his comment from the opening faculty meeting of 18 August when the plan was discussed, that it is necessary for Mines to change if it is to achieve greater levels of visibility and responsibility in the national and international arenas. Although such change would necessitate new strategies and entrepreneurial initiatives, Trefny reassured the community that his intention and that of the Board of Trustees is "to strengthen all aspects of the school, including traditional graduate studies and research." Noting that Mines' current size might be

The band marches down Washington Street at Homecoming.

Honor Code for Mines

The Associated Students of the Colorado School of Mines passed the following Student Honor Code in March of 2003.

PREAMBLE. The students of Colorado School of Mines (Mines) have adopted the following Student Honor Code (Code) in order to establish a high standard of student behavior at Mines. The Code may only be amended through a student referendum supported by a majority vote of the Mines student body. Mines students shall be involved in the enforcement of the Code through their participation in the Student Judicial Panel.

CODE. Mines students believe it is our responsibility to promote and maintain high ethical standards in order to ensure our safety, welfare, and enjoyment of a successful learning environment. Each of us, under this Code, shall assume responsibility for our behavior in the area of academic integrity.

As a Mines student, I am expected to adhere to the highest standards of academic excellence and personal integrity regarding my schoolwork, exams, academic projects, and research endeavors. I will act honestly, responsibly, and above all, with honor and integrity in all aspects of my academic endeavors at Mines. I will not misrepresent the work of others as my own, nor will I give or receive unauthorized assistance in the performance of academic coursework. I will conduct myself in an ethical manner in my use of the library, computing center, and all other school facilities and resources. By practicing these principles, I will strive to uphold the principles of integrity and academic excellence at Mines. I will not participate in or tolerate any form of discrimination or mistreatment of another individual.

a limiting factor, he emphasized that the school would always remain relatively small though modest growth over a ten-year period could "bring greater credibility to Mines as a major player among engineering colleges." To coordinate and facilitate the change necessary, a Strategic Planning Task Force was mobilized. While the question of the relationship between graduate and undergraduate education at Mines has often been debated, it is important to realize that because of economic vagaries, planning is, and must be, a continuous effort in higher education, even if it means revisiting, and learning from, the past, as well as attempting to gauge the future.

Another development with significant implications for the entire Mines community was a reunification agreement between the school and the alumni association that hopefully will bring the two entities back into a closer and more fruitful relationship. Key components of the agreement are the following:

Kick 'em high!

Marv Kay

Even a quick look at the career of Marv Kay *(pictured at left)* shows a man of many hats—most of them worn while he was active at Mines (forty-seven years) as student, athlete, coach, faculty member, and director of athletics, and with Golden, as city council member and mayor—not to mention memberships and chairmanships on innumerable community boards and organizations and his activities supporting the Mines Alumni Association.

Kay was, in a sense, born at Mines; that is, his father was a junior at the time and an All-American member of the undefeated 1939 football team. Kay followed in his father's footsteps, enrolling at Mines and participating in football and wrestling. Not only was he voted the Outstanding Sophomore and Junior Athlete at Mines, he was also elected to the Small College All-American football team. His skills in football, moreover, earned him a free-agent year with the Denver Broncos. Following his graduation (Mining Engineering) in 1963 and a stint in the Army Engineer Corps as a first lieutenant, he worked for a year with Ingersoll-Rand. But a career in engineering was not to develop for the young Kay.

At the urging of Fritz Bennecke, Mines football coach, Kay took a position as head freshman football coach and head swimming coach at his alma mater—a decision he has never regretted. In 1969, he was named head football coach, serving in that position until 1995 when he was promoted to director of athletics. From 1985, he held the rank of professor of physical education and athletics. Following two years on the Golden City Council, Kay served two terms as mayor—after which he continued on the council for another ten years. Kay's coaching honors include National Association of Intercollegiate Athletics 1975 District Coach of the Year; Rocky Mountain Athletic Conference 1975 and 1979 District Coach of the Year; and American Football Coaches Association 1975 and 1979 Coach of the Year.

Kay retired from his position as director of athletics in 2004. An outstanding example of the Mines spirit, Kay himself summed up his career best: "I'm very happy. I'm healthy. I've been blessed with a wonderful wife, family, and grandchildren. I've had the opportunity to live in Golden and fulfill a career at Mines. It's been a perfect fit for me. What more could a Miner ask for?"

- The Office of Alumni Relations (OAR) will be created within the Mines administration.
- The Alumni Association will retain its independent corporate identity and its Board of Directors and officers.
- The Director of Alumni Relations will be employed by Mines and report to the Mines president as well as to the Alumni Association's Board of Directors.
- The Mines president and the association's Board of Directors will develop the OAR's budget jointly.
- Funding for all alumni relations programs will be included in the president's budget.

Epilogue

- Alumni Association endowments shall continue to be the fiduciary responsibility of its Board of Directors.

"We did not take this decision lightly," noted Alan Mencin, CSMAA treasurer, "and it was one that we knew could be perilous. We felt that this was the best choice because our duty is to the alumni and the school" (MM Summer 2003). The Alumni Association is at last returning home, and both alumni and school should be the better for it.

Dr. Paul Sharp, a former president of Drake University, speaking at a freshman convocation some years ago, said in a moment of Thoreauvian simplicity, "A university should be three things: intellect, intellect, intellect." Though this work has focused extensively on change at Mines, some aspects of a collegiate institution should remain constant. The sharpening of minds is surely one. The tools used may differ, but the goal is the same: individuals who are prepared to venture into what the engineer-poet Alan Dugan referred to as "the daily accident" and work to make the world less menacing and less accidental. Training should be education, and education training. The two go hand in hand. Only a shallow mind would see them in opposition.

Although this telling of the Mines story is finished, the larger Mines story is not. Having established itself as a major player in the education and research

The Orediggers move the ball in the Homecoming game.

demanded by the mineral industries of the world, Mines may justifiably look ahead with pride and confidence to a continuation and an expansion of its mission. For many, the history of an educational institution is simply a record of past events and dead people. True enough, but it is more than that. Just as the imagination is guided by the past, so too is it challenged by the future. Every student who has studied at Mines, every professor who has taught there, and every administrator, staff member, and trustee who has worked to guide and support the school in its educational mission, past or present, is as much a part of the Mines story as will be those who follow. Objectively speaking, the Mines spirit may be no different from the spirit of any other college or university. Those who experience it and contribute to it, however, know the satisfaction of being part of an institution based on hard work and achievement that has its own creative power.

Born at the right place at the right time and supported and strengthened by its many constituencies, Mines continues to turn vision into reality, as it extends its reach

from the

Rocky Mountains to the World.

Go, MINES!

The burros of Mines reflect the sunlight.

The prize-winning Center for Technology and Learning Media.

The Union.

Mount Zion looks down on the Volk Gymnasium.

Epilogue

Kafadar Commons.

Stratton Hall.

Looking toward the Union.

Though not highly rated nationally for its food, the cafeteria provides nourishment and fellowship.

A class of seniors hears a lecture on military history in the well-appointed lecture room in Hill Hall.

Rocky Mountains to the World: A History of the Colorado School of Mines

A Homecoming game touchdown!

With the ball Stephen Bahl moves to the basket.

He's a "big digger" for sure.

The long haul for the long (cardboard) boats.

Doug Baldwin, in a farewell gesture and with advice from Mines Magazine editor Maureen Keller, took it upon himself to publish the last issue of the Prospector (2000–01).

The cardboard boat race is on at nearby Clear Creek.

Epilogue

President John Trefny welcomes his turn in leading Mines toward new challenges and new achievements.

The 2004 Board of Trustees, like their predecessors—and like the "M"—watch over Mines. Seated left to right: Dr. Michael Nyikos, F. Steven Mooney '56, and Terrance Tschatschula. Standing left to right: Hugh Evans '49, Kale Franz '04, Karen Krug '84, David Wagner, and John Coors '77.

The Thursday lunch bunch (alumni) at the Buffalo Rose. Good food and good stories.

Marvin the Miner, a new Mines mascot, named in honor of Marv Kay.

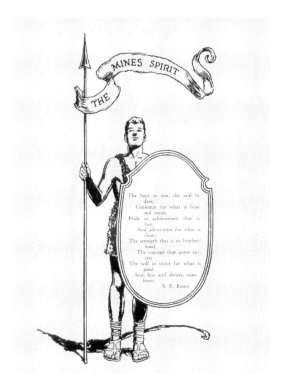

Perhaps the Mines spirit cannot be defined or personified, but the class of 1930 provided their view of it.

217

Bibliographical Sources

Among the many available sources used in the preparation of this history of Mines, the following were the most helpful.

The Alumni Association

The Alumni Association has complete sets of *Mines Magazine* and *Prospector* (school yearbook), as well as a broad, though not complete, set of the *Oredigger* (school newspaper). Also available are many photographs depicting alumni events and activities.

Arthur Lakes Library

The Arthur Lakes Library has complete sets of the preceding items, along with microfiche of the *Transcript* (Golden newspaper) from its beginnings. The Russell L. and Lyn Wood Mining History Archive has a wealth of materials, including unpublished papers, documents, letters, and newspaper clipping files relating to the school; school catalogs (from the founding of the school to the present); oral histories; all of the presidential biennial and annual reports; a full set of the *Oredigger*, a full set of *Mines Quarterly* (now *Quarterly Review*), a number of additional school publications; a wide collection of books and periodicals on mining and metallurgy; and countless photographs (partially catalogued) of people, buildings, and campus activities. In short, this archive is the place to begin any study of Mines.

Files of the Office of Academic Affairs

Because of their volume and cramped storage space, these files are not as easy to use as they might be, but the office staff is more than willing to assist in finding and using the materials. These files contain virtually a day-by-day catalog of all the administrative activities in the operation of the school. Particularly valuable are academic plans, North Central Association reports, American Association of University Professors reports, American Board of Engineering and Technology reports, and other evaluative reports from various groups.

Academic departments, athletics, special programs

The academic departments have written histories and current records of their activities—the latter through newsletters, annual reports, and other forms of communication. Numbers 4 of 1996 and numbers 1, 3, and 4 of 1997 of the *Quarterly Review*, have excellent histories of eight departments. The Athletic Department has considerable historic material, ranging from records and statistics to photographs and various published materials. Programs such as EPICS, the Guy T. McBride Honors Program in Public Affairs for Engineers, musical and dramatic organizations, and SPACE have their own brochures and publicity materials.

Mines Magazine, Mines Today, the *Oredigger,* and *Prospector*

These publications are full of excellent material pertinent to the history of Mines. *Mines Magazine* is, of course, a "gold mine" of information. This magazine, from its inception to the present, provides a wealth of historical information on the school. *Mines Today* (no longer published) also contains good articles covering various aspects of Mines. The *Oredigger*, generally well balanced and well written, provides the student view of the school and its various activities, academic and social. One can get a good picture of Mines from a thorough reading of this paper. *Prospector*, unfortunately no longer being published, began in 1913 and continued until

2001. Like all yearbooks, it is filled with pictures of individuals, groups, and activities, as well as short articles. Taken together, these four publications cover the history of Mines thoroughly and, for the most part, objectively.

Newspapers

Little from the past or the present dealing with Mines was, or is, missed by the *Transcript*, the *Denver Post*, or the *Rocky Mountain News*. The *Transcript* has a kinship with Mines that dates back to the school's founding, and virtually every edition covers some aspect of Mines' operation. The *Post* and the *News*, both excellent newspapers, have given Mines its share of coverage, particularly regarding some of the more sensational incidents in Mines' history.

Histories, published and unpublished

The only book-length, formally published general history of Mines is Jesse Morgan's *A World School: The Colorado School of Mines*, Sage Books, 1955 (available in Arthur Lakes Library). It is a highly readable and informative potpourri of historical facts, anecdotes, traditions, and photographs that presents Mines from a number of angles.

Other general histories include Regis Chauvenet's *History of the Colorado School of Mines: Origin and Early Years*, 1920 (available in Arthur Lakes Library); Montgomery Budd's *Colorado and Its School of Mines*, 1929, appearing in MM from April 1929 to July 1930; Mary Hoyt's "A Short History of the Colorado School of Mines," 1949, appearing in MM June 1949; Den Galbraith's "Give 'em Hell, Mines," appearing in MM February 1974; James T. Smith's "Early History of the Colorado School of Mines," *Quarterly Review* (Colorado School of Mines), Vol. 100, No. 1, 2000; Robert Sorgenfrei's "The Colorado School of Mines: Its Founding and Early Years, 1874–1902" *Quarterly Review* (Colorado School of Mines), Vol. 100, No. 2000; and a small paperback, *Homecoming 1973, CSM Centennial*, containing a short, concise history of the school up to 1973.

A Century of Women at Mines is a full-length paperbound book published at Mines in 1999 (funded by the Chevron Corporation) that, through a statistical narrative and anecdotes written and edited by a panel of authors, provides excellent coverage of the experience of women at Mines from the late nineteenth century to the present. Containing an excellent bibliography, it is essential reading for any study of Mines history.

An unpublished book-length history by Professor Emeritus Frank Hadsell, *History of the Colorado School of Mines by Divisions (Booms and Busts)*, 1997, provides an insider's view of Mines, particularly of the departments of Geophysics and Geology. Written with touches of humor, the book contains an abundance of information that complements the above histories.

Professor Emeritus of Geology Robert Weimer's *125 Years of Earth Science at CSM: Lessons for the Future*, a CSM Senate Distinguished Lecture (December 1999) presents a brief but informative historical overview of Mines and looks to Mines' future. Weimer has also written informative articles for MM, as have Professor Emeritus of Geology John Haun, Robert Sorgenfrei, and Patricia Petty. Fritz Brennecke wrote many articles for MM and the *Transcript* that cover Mines athletics and general history.

Anecdotal conversations

Needless to say, talking with faculty and administrators who have been at Mines for some years and alumni of various ages yields the kind of anecdotal information that is generally not found in books or articles. Usually objective, though sometimes not, such information cuts through to the core of a school's struggles and accomplishments, giving it a life that a single author cannot.

Note

At this time, it is not clear where materials used in this work will be filed. Certainly, they will be organized and kept in a specific location and accessible to anyone interested in the history of Mines. Virtually all of the material was "mined" from the above sources. There is enough unused material for another history, perhaps with a more specifically focused approach.

Index

Abu Dhabi, 11, 209
Accrediting Board of Engineering and Technology, 86
act to establish a school of mine, 21
Adelbert College, 36
Advanced Steel Processing and Products Research Center, 190
Agitation Week, 177
Agricola, Georgius, 31-32, 75
Alderson Hall, 10-11, 90, 134, 140, 195, 199
Alderson, Victor C., 9-10, 43, 46, 49-57, 59, 62, 63, 67-68, 73, 75, 78, 80, 82-85, 90-91, 134
Alexander, William, 129
Allen, Carl A., 56
Allen, Reverend Ethan, 17
Alpha Delta, 168, 182
Alpha Tau Omega, 182
Alumni Advisory Council, 159
Alumni Association, 6, 9-11, 36, 53, 81, 84, 89, 96, 98, 106-107, 118, 120, 123, 165, 177, 193, 323-213, 218
Alumni Council, 118
Alumni Diamond Jubilee, 118
Alumni Endowment Fund, 118
American Association of Engineers, 80
American Association of University Professors (AAUP), 11, 85-87, 89-90, 149, 165-167, 177, 190, 224
American Dream, the, 13, 95
American individualism, 95
American Institute of Mining Engineering, 99
American Society of Engineering Education, 128
Amsbary, William, 23
Anaconda Copper Company, 51
Ancell, Ken, 151
Ansell, George, 11, 92, 185-187, 189-192, 198-201, 204, 207
Armour Institute of Technology, 49

Army Corps of Engineers, 123, 162
Arthur Lakes Library, 6, 10, 137, 147, 194, 218-219
Assay Laboratory, 9, 33, 35, 39, 51, 105, 153
Assaying, 21, 24, 32-33, 47, 92
Atlantic Richfield Company, 192
Avery, Nathan, 206

Bahl, Stephen, 206, 217
Baldwin, Douglas, 6, 216
Ball, Max W., 90, 93, 103, 107
Baltimore, 17
Band, 80, 85, 91, 182, 186, 188, 196, 198, 210
Banning, James H., 194, 197
Barbarians ("Barbs"), 79, 174, 182
Barbeque, 44, 66, 105
Barron, Irving, 59
baseball, 21, 24, 44-45, 58, 63-64, 69, 76, 80-81, 98
basketball, 44, 47, 54-56, 58, 69, 81, 91, 108, 111, 151, 179, 180, 207
Beadle, George, 46
Bell, Charles N., 96
Ben H. Parker Student Center, 63, 149-150, 153, 197, 199, 204
Benchmark, 175
Bergakademie School of Mines, 22
Berthoud Hall, 10-11, 16, 100, 110-111, 114, 134, 199, 214
Berthoud, Edward L., 16, 20, 23
Beta Theta Pi, 83, 110, 135, 164, 168, 170, 182, 209
Bickert, Theodore, 11, 185, 202-204, 207
Blackhawk, 16
Blaster, 209
Blasters, 62, 109
Blaurock, Carl, 107
Bloch, Ivan, 76
Blue Key, 63, 79, 102, 137, 177

Board of Trustees, 15-16, 23-24, 26-27, 29, 34, 37-38, 42, 55-57, 61-62, 64-65, 67-68, 75, 78, 81-83, 85, 89-93, 98, 102, 110, 113, 124,129, 131, 133-134, 136, 143, 146, 150, 152, 159, 165-166, 170, 174, 177, 189, 193, 198, 201, 206, 217
Board, Gregory, 24
book miners, 23
Borgmann Report, 156
Borgmann, Carl W., 156
Borthick, Jacquiline, 135
Boston and Colorado Smelter, 16
Bowman, Frank C., 118
Boyd, James, 150
Braden, C. H. C., 102
Bradford Hall, 10, 119, 135
Bradford, A. Hartwell, 135
Breit, Kathy, 6
Brennecke, Fritz, 44-45, 47, 219
Brooklyn Bridge, 129
Brooks Athletic Field, 79, 81, 106, 135, 178, 207
Brooks, Ralph D., 81
Brown Hall, 11, 175
Brown University, 15-16
Brown, George R., 175
Brown, James, 190
Brown, Martin, 103
Brown, Walter, 63
Buckley Air Field, 121
Bull, Rex, 172-173
Bulletin of the American Association of University Professors, 83, 166
Burdick, H. D., 130
Burns, Norman, 142-143
Butler, G. Montegue, 56

Caldwell Hall, 162-163, 179
Caldwell, Florence, 9, 36-37, 44-45
California, 13-14, 58, 73, 86, 131, 135-136, 149, 152
campus culture study, 194-196
Carnegie, Andrew, 31
Carson, Clyde, 103

CCHE Excellence Award, 192, 203
Centennial, 24, 145, 163, 166-167, 169, 176, 185, 204, 219
Center for Engineering Education, 209
Center for Geoscience Computing, 190
Center for Technology and Learning Media, 11, 210
Center for Wave Phenomena, 190, 208
Central City Register, 18
Central City, 14, 18, 24
Challenge Grant, 192-193
Channing, William Ellery, 49
Chapman, Carol, 6
Chauvenet Hall, 153, 206
Chauvenet, Regis, 9, 29, 31-42, 44, 47, 50, 56, 61, 91, 155, 219
Chemical Engineering and Petroleum Refining, 187, 191, 203
Chemistry, 21-24, 31-32, 34, 38, 40, 42, 50, 54, 61, 78, 91, 99, 102, 121-122, 125-126, 130-131, 133, 143, 147, 149, 150, 154, 159, 170, 176, 191
Chesbro, Steve, 209
Chevron Corporation, 219
Childs, Orlo E., 10, 144, 147, 152-153, 155, 159, 162, 171, 192
Chorus, 176, 188, 198
Civil Engineering, 18, 22-23, 28, 92, 130, 203
Civil War, 16, 73, 148
Clear Creek Canyon, 44, 46, 105, 132, 164
Cold War, 117, 125, 178, 185
Collburn, C. Lorimer, 118
College of William and Mary, 180
College Union, 47
Colorado A & M, 46, 55
Colorado Center for Advanced Ceramics, 190
Colorado College, 46, 55, 91, 109, 127

Index

Colorado Commission on Higher Education, 144
Colorado House Committee on Education, 165
Colorado Mining Association, 125-126
Colorado School of Mines Foundation, 10, 81, 118, 133, 193, 200
Colorado School of Mines Quarterly, 26
Colorado School of Mines Research Foundation, 159
Colorado School of Mines, 6, 8, 10, 13, 16, 25-26, 49, 59, 77, 93, 96, 100, 104, 115, 124, 130, 132, 134, 140-141, 144, 151, 157, 167, 193, 196, 198, 200, 211, 219
Colorado Springs, 25, 37, 73, 127
Colorado State University, 156
Colorado Territory, 9, 15-16, 21
Colorado Transcript, 8, 19, 25
Columbia College, 20
Columbia School of Mines, 22
Conger, Harry, 200
Coolbaugh Hall, 11, 125, 133-134, 199
Coolbaugh, David, 6, 109
Coolbaugh, Melville, 10, 90-93, 96-99, 101-105, 108-110, 112-114, 118-119, 122, 134
Coors Brewery, 21, 27, 141, 200
Coors, John, 217
Coors, William K., 172, 200
Corps of Cadets, 21
Cousin Jack miners, 16
Cripple Creek, 14, 22, 37, 102, 114
Culver, Richard, 192
curriculum, 15, 24, 27-28, 32-33, 41, 54-56, 72, 77, 79, 90, 92-93, 101-102, 105-106, 122-123, 126, 128, 130-131, 134, 140, 142-143, 145, 150, 159, 162, 165-166, 170, 174-175, 186-187, 190-192, 194, 197-198, 200-201, 203

D. Allan Bromley, 206
Dames Club, 124
Daniel, Steve, 176
Darden, Jim, 152
Davis, C. C., 23
Davis, Ninetta, 37
Deadwood, 14
DeFord, Ronald, 106
degrees offered, 10, 101, 105, 118, 125
Denver Post, 161, 219
Denver Republican, 27, 39
Denver, 9, 15, 19-20, 22-25, 32-33, 40, 46, 59-62, 72, 89, 91, 93, 114, 127, 131, 192, 196, 200, 206, 212
departmental enrollments, 191
Depression, the, 93, 95-98, 101, 104, 110, 117-118
Dickinson, Donald, 6
dormitories, 10, 71, 81, 105, 119, 121, 125, 135, 147, 158, 163, 179, 210
Dotson, Major G. C., 75
Drake, E. L., 102
Dynamiters, 62

Easley, Nancy, 117, 135, 185
Eckley, Timothy, 6
Economics, 6, 11, 23, 31, 102, 130, 143, 150, 159, 170, 186, 191, 193, 197
Edgar Mine, 10, 132, 191
Educational Testing Service, 130
Einstein, Albert, 41, 206
Eisenach, Edward, 129
Electrical Engineering, 35-36, 41, 50, 130
Eliot, Charles William, 47
Ellsworth, "Shorty", 47, 57, 59
Emerson, Ralph Waldo, 201
Encyclopedia of the Korean War, 123
Engineering and Mining Journal, 38, 42, 65, 86, 93
Engineering Day, 10
Engineering Practices Introductory Course Sequence (EPICS), 6, 192, 218
Engineering, 9-11, 16, 18, 21-23, 28, 33, 35-36, 41, 49-51, 53-55, 75, 78-81, 86, 90, 92-93, 96, 98-99, 101-102, 104-105, 107, 111-112, 120-122, 125-131, 134, 139-143, 145, 147-150, 153-154, 156-157, 159, 161-163, 165-167, 169-172, 175, 178, 185, 187, 190-193, 196-203, 206, 208-209, 211-212
Engineers Company of the National Guard of Colorado, 54, 73
Engineers Council for Professional Development, 153
entrance requirements, 39, 102, 122, 131, 147
Environmental Science, 190-191, 197
Erisman, Frank, 201
Evans, Hugh, 136, 217
Exemplary institution, 11, 206, 209

Experimental Plant, 35, 52, 54, 57, 80, 90, 112, 123, 133

faculty development, 175, 193
Faculty Senate, 11, 165, 187, 190
faculty tenure policies, 149, 190
faculty unrest, 10, 56, 188
Federal Housing Administration, 119
Feighny, Kathleen, 6
Fertig, Colonel Wendell W., 81, 120, 123
Field House, 10-11, 110, 119
Fischer, Roland B., 32-33
Fisk, Barbara, 179
Flu Epidemic of 1918, 72-73
Flute Choir, 188
football, 9-10, 39, 44-47, 55, 58-61, 77, 79, 81, 84, 88, 103, 107-109, 111, 127, 129, 168, 178-179, 188, 193, 207, 210, 212
Fowler, Eugene, 59
Fowler, Vera, 180
Franco Wyoming Oil Company, 97
Franz, Kale, 217
fraternities, 44, 53, 61, 71-72, 79-80, 89, 105, 112, 126, 144, 163-164, 168, 177, 179-182
Frederick Jackson Turner, 139
Freshman Orientation Week, 177
Future Graduate, 173

Galbraith, Den, 15, 41-42, 45, 57, 59, 219
Gary, James, 159
Gauntlet, the, 10, 44, 113, 126, 137, 144
General Theological Seminary of New York, 15
Geological Museum, 100
Geology, 6-7, 18, 21-23, 43, 50, 52, 54-56, 79-80, 91, 95, 98-99, 102, 104, 110, 119, 122, 129, 131-133, 145, 147-149, 152, 170, 191, 203, 219
Geophysics, 6, 10, 93, 98, 103-104, 109, 122, 125, 129, 133, 139, 146-148, 191, 200, 203, 219
George S. Ansell Department of Metallurgical Engineering and Materials Science, 11, 201
George, Clarissa, 182
Gilot, Francois, 161
Glaze, Ralph, 88
Glee Club, 57, 60, 80, 87, 91, 102, 188
Global Systems and Cultures Department, 193

gold and silver production in Colorado, 14-15, 22, 31-32, 92, 99-101, 132, 196
gold, discovery of, 9, 13-14, 16, 196
Golden Arches, 139
Golden City Park, 73
Golden City, 14, 18-19
Golden Globe, 25, 68
Golden, John, 187
Golden, 6-7, 9, 14-22, 24-27, 29, 34, 40-42, 46, 49, 61-65, 68, 71-73, 77, 83, 86, 103, 112-113, 119, 135, 141, 153, 171, 209, 212, 218
Gonzales, Suzann, 182
Gorbachev, Mikhail, 185
Goulding, Genevieve, 6
graduate education at Mines, 146, 157, 159
graduate education following World War I, 145
graduation requirements, 158
Graves, Ramona, 197
Green Center, 11, 153, 155, 176, 199
Green, Cecil, 155, 200
Green, Ida, 155
Gregory, John, 14
Grewcock, Bruce, 206
Grinter, L. E., 128
Griswold, William, 131
Guggenheim Hall, 9, 52-53, 61, 80, 117, 136, 146-147, 153, 171, 180, 185, 188, 206
Guggenheim, Meyer, 31, 52
Guggenheim, Simon, 9, 52
Gunnison Tribune, 40
Guy T. McBride Honors Program in Public Affairs for Engineers, 193, 218
Gymnasium, the, 9, 11, 34-35, 43, 47, 54-56, 61, 80-81, 84, 145, 214

Hadsell, Frank, 6, 104, 172, 183, 219
Hahn, Chuck, 178
Haldane, William G., 10, 56, 61-62, 69
Hale, Albert, 9, 21, 27-29, 33, 42
Hall of Advanced Chemistry, 80
Hall of Chemistry, 27, 80
Hall of Engineering, 9, 35, 80
Hamilton, Jerri, 100
Hanley, William, 59
Harsh, Levi, 19, 21
Harvard College (University), 20, 29, 47, 49, 51, 93, 96, 124
Harvard Magazine, 47
Hathaway, C. M., 127-128
Haun, John, 6, 80, 158-159, 170, 172, 219
Hawkins, G. A., 141-142

221

Heikel, Ramona, 168
Heiland, C. A., 93, 103-104
Hennebach, Ralph, 97, 192-193
Henry, C. L., 206
Heraclitus, 185
Hewlett, Doc, 112
High Grade, 176
Hill Hall, 11, 147, 199, 201, 205, 215
Hill, Benjamin, 93, 102
Hill, Charles, 107
Hill, Louis C., 35
Hill, Nathaniel, 16, 23148
Hilltrop, Joni, 182
Hogan, John, 176
Homecoming, 10, 44, 63, 81, 86, 91, 135, 144, 164-166, 181, 209-210, 213, 216, 219
Honor Code, 211
Hoover, Herbert, 23, 95, 107, 224
Horizon Plan, the, 133, 148
Hoskins, A. J., 56
Hoyt, Mary E., 87, 219
Hull, Charles, 97
Humanities, 23, 122, 143, 153, 165, 170, 173, 176, 188, 191-194, 224
Hundred Year Horizon, 131
Hutt, George, 51-52
Hynes, Fred, 112

Idaho Springs, 14, 132
Independence Mine, 37
Integral Club, 47, 84
International Ground Water Modeling Center, 190
Istanbul Technical University, 199

Jackson, George, 14
Janzen, Howard, 209
Jarvis Hall, 9, 15-22, 26, 45, 73, 100
Jarvis, George, 17
Jarvis, Mrs. George, 17
Jefferson County Republican, 112
Johnson, H. A., 14
Johnson, Harlan, 7, 100
Johnson, Kristina, 206
Johnston, Dave, 46
Johnston, William, 59
Junior Affiliated Society of the American Institute of Mining and Metallurgical Engineers, 80

Kafadar Commons, 46, 198, 215
Kafadar, Ahmed, 193
Kammerzell, Don, 166
Kappa Sigma, 80, 111, 168-169, 172, 179, 182
Kay, Marvin, 212, 217
Keller, Maureen, 6

Kennedy, George H., 170, 172
Kennedy, John, 178
Kennedy, Robert, 178
Kent State University riots, 161
Kerr, Clark, 79
King, Martin Luther, 178
Kircher, Rex E., 135
Korean War, 123, 126, 128, 130, 178
Krug, Karen, 217
Kuhn, Truman, 140, 163

laissez-faire economics, 95
Lakes, Arthur, 22-23, 28, 36, 38
Lamb, Richard D., 172
Latin American Center, 193
Laubach, Winton, 166, 190
Law, Dan, 166
Lawson, Ted, 103
Leadville, 25, 119
Leidich, Pete, 112
Lessing, Lawrence, 127
Liberal Arts and International Studies Division, 106, 224
Library, 6, 8-10, 17, 22, 29, 34-35, 45, 47, 52-53, 121, 137, 147, 193-194, 200, 211, 218-219
Loveland, W. A. H., 15, 21, 23
Lulea University, 199
Lynch, Colleen, 180

M, 9-10, 44, 61, 63, 67, 110, 152-153, 157, 177, 198-199, 202, 214, 217
Madden, Lloyd, 107, 110
Mageneau, William, 34
Main Hall, 51
Mallett, E. J., 9, 23-24
Manifest Destiny, 13, 31
Marvin the Miner, 217
Mason, John, 109
Master Plan, 1982-88 174
Mathematics, 6, 18, 23, 33, 40, 49-50, 55-56, 104, 130-131, 143, 147, 149-150, 154, 159, 166, 191, 193
Matthews Hall, 9, 15, 17-18, 20-22, 24, 26, 73
McBride Honors Program, 188, 191, 193, 218, 224
McBride, Guy T., 10, 162-163, 165, 167, 170, 172-177, 185, 190-192, 199
McDermut, Grace, 36-37
McFall, George, 166
McKee, Robert III, 206
McNeil, Charlie, 177
Mechanical Engineering, 92, 130
Mencin, Ala, 213
Metallurgical and Chemical Engineering, 52

Metallurgy, 11, 18, 20-22, 24, 28, 32-33, 41, 43, 50-51, 54-55, 73, 101-102, 122, 129, 133, 139, 146-148, 200-201, 218
Metals Exploration Company, 91
Meyer Hall, 10, 148
Meyer, Paul, 33, 40-41, 50, 153
Mickey Mouse, 171, 180-181
Middleton, Nigel, 208
Middleton, William B., 9, 28-29, 36
military drill, 19, 54, 72-73, 75, 111, 113, 131, 161, 165
Military Science Department, 73, 113, 192, 198
Miller, Maureen, 6, 216
Mineral Economics, 150, 159, 197
Mines Magazine, 6, 8, 53, 89, 106, 118, 216, 218
Mines Park, 134, 199, 204-205, 210
Mines spirit, 46, 57, 79, 84, 91, 96, 107, 114, 119, 151, 178-179, 193, 210, 212-213, 217
Mines Sports Hall of Fame, 152
Mining American, 67
Mining and Scientific Press, 67, 86
mining education in the United States, 23-24, 53-54, 86, 90, 132, 196
Mining Engineer song, 87, 106, 178
mining engineers, 22-23, 34, 41-42, 49, 52-55, 78, 86, 90, 92-93, 119-120, 122, 132, 149, 178, 191, 196, 200, 212
mining industry of Colorado, 16, 20, 53, 125
Mining University of Leoben, 199
Mining, 9, 13-26, 31, 32, 34, 37, 41-42, 49-56, 52, 65, 74, 78-80, 86, 89-93, 99, 101-102, 105, 109, 119-120, 122, 124-126, 129-130, 132-133, 144, 147-149, 153, 166, 174, 190, 196, 200, 202-203, 209, 218
Missouri River, 23
Money, Nancy, 165
Monterey Institute of Technology, 199
Montoya, Eloise, 180
Mooney, F. Steven, 217
Morgan, Jesse R., 101-102, 104, 158, 196, 219
Moss, Milton, 9, 20, 25-26, 28
Mount Zion, 44, 63

Mouseketeers, 139
Mu Epsilon Tau, 80
Music, 15, 40, 147, 161, 188, 201

National Collegiate Athletic Association, 109, 179
National Defense Act of 1920, 73
National Endowment for the Humanities, 191, 224
National Science Foundation, 146, 208
Nelson, J. W., 141-142
Newcomen Society, 167
Newman, Henry, 71
Newsweek, 11, 173
North Central Association of Colleges and Secondary Schools, 86, 142, 153, 169, 172
Nyikos, Michael, 217

Office of Alumni Relations, 11, 212
Ohio Wesleyan University, 36
Old Chemistry, 42, 134, 147
Olds, Barbara, 6
Oredigger, 6, 80, 87, 97, 106-107, 136-137, 144, 166, 170-171, 178, 194, 198, 210, 218
Owens, Bill, 206

Palmer, Charles S., 9, 39, 42, 44
Palmer, Irving A., 99, 101
Palmer, Robert S., 125-126
Parker, Ben H., 10, 118-119, 121-122, 124-125, 129, 152
Parker, Charles O., 33
Parmalee, Howard C., 10, 62, 64-65, 69, 82, 85
Patton, Horace B., 43
Paul, J. Marshall, 23
Pavelich, Michael, 192
Pegis, Anton G., 6, 163
Pennsylvania State University, 112, 178, 224
Peterson, Ralph, 206
Petroleum, 10, 79-81, 93, 98-99, 101-102, 105, 109, 119, 121-123, 125-126, 129, 133-135, 139, 147-149, 176, 181, 187, 191, 193, 197, 200, 203, 208-209
Phi Beta Kappa, 53, 180
Philipose, Tom, 191
Phillips, William Battle, 10, 61-62, 69, 83
Physics, 10, 19, 35-36, 38, 40, 50, 54, 99, 102, 104, 121-122, 130-131, 143, 147, 149-150, 153-154, 159, 188, 191, 207
Pi Beta Phi, 182
pick and pan prospectors, 14

Index

Pitts, J. Douglas, 162
Plank, William B., 105
Platte River, 13
Powers, David, 6, 200
Price, Collette, 182
Prospector Park, 134
Prospector, 6, 74, 80, 216, 218
Puttu.ck, Leslie, 180

Radio Station 9XAI, 74, 80
Randall Hall, 147
Randall, Bishop George, 9, 14-19, 22, 28, 147, 186, 206
Rath, Charles M., 96
Read, Thomas, 23
Rensselaer Polytechnic Institute, 185
Reserve Officers Training Corps (ROTC), 10-11, 71-75, 82, 111-112, 115, 120, 123, 126, 144-145, 154, 161-162, 165-166, 175, 177, 179, 198
Reservoir Characterization Project, 190, 208
Resource Fund, 172
RESOURCES Campaign, 200
Reybold, Major General Eugene, 123
Rhodes, Frank H. T., 139
Rickard, T. A., 93
Rico News, 51
Roaring Twenties, 78, 117
Roberts, Claudia, 182
Robotics Automated Welding Laboratory, 190
Rockefeller, John D., 31
Rocky Mountain Athletic Conference, 9, 46-47, 54-55, 58-60, 69, 76, 108, 151, 180, 207, 212
Rocky Mountain News, 18, 20, 219
Roosevelt, Franklin Delano, 95
Roosevelt, Theodore, 45-46
Royal School of Mines, 24, 52
Ryan, Lynn, 165

Sadtler, Benjamin, 33
Salzer, George W., 107
Sankovitz, James, 179
Sayre, Robert H., 93
Schlechten, A. W., 163
Schmidt, Kaitlin, 6
Schneider, George W., 56
Scholastic Aptitude Test (SAT), 131, 149, 155, 161, 172, 194, 204
School traditions, 8, 43-44, 57, 91, 101, 105, 126-127, 129, 136-137, 144, 177, 179, 195-196, 202, 219
Schwartzberg, Fred, 185

Scott, George, 108
Scott, Sir Walter, 170
Seeman, A. K., 106
Senior Challenge, The, 113
Senior Day, 62-65, 99, 101, 111, 113-115, 127, 136-137, 171
Sharp, Paul, 213
Sheffield Scientific School of New Haven, 20
Sigma Alpha Epsilon, 80, 111, 173, 182
Sigma Gamma Epsilon, 80
Sigma Nu, 80, 110, 168, 172, 180, 182
Sigma Phi Epsilon, 113, 168, 173, 182
silver diplomas, 10, 97, 105, 107
Simpson, William S., 89
Sir Billy (goat), 61
Sloan, Dendy, 187
Slothower, Ben, 102
Smiley, William H., 93
Smith, A. C., 56
Smith, Albert, 72-73
Smith, C. E., 56
Smith, Captain James T., 24, 26, 219
Smith, Patricia, 182
Social Darwinism, 31
Society of Women Engineers, 198
Sociology, 23
Socrates, 201
softball, 127, 180
Sorgenfrei, Robert, 6, 29, 41, 219
sororities, 44, 168, 180-182, 210
South Dakota School of Mines, 49, 74
Soviet Union, 110, 117, 147, 167
special courses, 22, 24, 33, 36, 41, 192
Special Programs and Continuing Education, 181, 203
special students, 9, 24, 28, 33, 36, 41
Spock, Dr., 139
St. John's of the Wilderness, 15
St. Louis Democrat, 20
Staff, Greg, 164
Steele, Brenda, 182
Steinhauer Fieldhouse, 10-11 199
Steinhauer, Frederick, 10, 57, 110
Steinhauer, Gurnett, 163
Stetson, the, 44, 179, 196
Stevens, Wallace, 76
Stevinson, Greg, 206
Stewardship of the Earth, 190, 202

Stockton, Robert, 36, 118
Strategic Planning Task Force, 208-209, 211
Stratton Hall, 9, 11, 37, 51, 66, 80, 135, 199, 206, 215
Stratton, Winfield, 9, 37, 51
Student Army Training Corps (SATC), 71, 73
student housing, 119, 121, 123, 125, 134-135, 147, 158, 163, 179, 199, 204-205, 210
student unrest, 9-10, 38-42, 92
Stute, Lori, 182
summer school, 51, 104, 113, 121, 123
Swartz, Dan, 178
swimming, 37, 47, 56, 88, 179-180, 212

Tau Beta Pi, 53, 80, 114, 128
Terman Report, 156
Terman, Frederick, 156
Territorial Assembly, 18-21, 23
Territorial School of Mines, 8-9, 20, 23
Thaler, Sarah, 180
Theta Tau, 53, 80, 144
This Fascinating Oil Business, 103
Thistlewood, Jo Ann, 29
Thomas, Lester C., 134, 158
Thoreau, Henry David, 23, 95, 213
Thornton, Dan, 131
Thursday Lunch Bunch, 217
Times Record, 114
Transcript, 8, 19-21, 24-25, 29, 46, 114, 188, 218-219
Transforming Resources: The Campaign for Mines, 209
Trapenhagen, Frank W., 43
Trefny, John, 11, 102, 207-210, 217
Truly, Richard, 206
Tschatschula, Terrence, 217
Tug of War, 44, 105, 177, 183
Turner, Lee, 144

U. S. Engineers, 115th Regiment, 10, 73, 199
University of Colorado, 11, 42, 55, 58, 61, 63, 120, 135, 147, 156, 177, 186, 194, 198
University of Denver 46, 55-56, 59-61, 63, 81, 193, 198
University of Guanajuato, 199
University Schools of Colorado, 21

Van B, Lillian, 54
Van Lieu, Harriet, 182
Van Tuyl, F. M., 79-80, 104
Vanderbilt, Cornelius, 31

Vanderwilt, John W, 10, 124-126, 129, 131, 136, 140, 142, 145, 148, 150, 152-153
Vietnam War, 144, 148, 161, 167, 196
Vivian, John, 68
Volk Gymnasium, 11, 81, 145, 214
Volk, Russell, 81, 129, 145

Wagenbach, Lorraine, 29
Wagner, David, 217
Waltman, W. D., 97
Ware, W. W., 23
Warren, Gaylord, 105
Watkins, Boyd, 151-152
Watson, Michael, 6
Way, William A., 93
Weaver Towers, 203
Weaver, Gaylord "Buck", 203
Weimer, Robert, 6, 91-92, 219
Welch, Charles C., 15, 20
West Point of the Rockies, 111
West, George, 19-20, 25, 73
Whitman, Walt, 13
Wilkins, Erika, 6
Wiltse, M. A., 170, 172
Witkowsky, Dan, 196
Wolfe Hall, 15
women at Mines, 10, 28, 36-37, 92, 111, 135, 137, 150, 154, 162-163, 172, 177-180, 185, 190, 197-198, 204, 207, 210, 219
Women in Science, Engineering and Mathematics, 198
Works Progress Administration, 110
World War I, 42, 73, 110, 117, 119
World War II, 75-76, 91, 95, 98, 109, 117, 119-121, 123, 125-126, 145, 178, 182, 193
World, 20
Wright, Alpheus, 23
wrinkling, 64, 177
Wrobel, Sandy, 180
Wuensch, C. Erb, 87, 89

Yale University, 47, 51, 178, 206, 224
Yench, Jackson, 171
Yom Kippur War, 167

Zehr, Norm, 6, 123

About the Author

Dr. Eckley came to Mines in 1984 as head of the Department of Humanities and Social Studies. He held that position until 1991 when he became acting director of the Liberal Arts and International Studies Division. He also served as principal tutor of the McBride Honors Program from 1989 to 1991. Prior to coming to Mines, he was professor of English at Drake University for nineteen years, serving as chair of the department for fifteen of those years. He was Senior Fulbright Professor at the University of Ljubljana in Slovenia (1972–73) and at Cyril and Methodius University in Bulgaria (1981–82). In 1993–94, he was visiting professor at Bilkent University in Turkey.

Eckley's publications include two books on Southern novelists (Harriette Arnow and T. S. Stribling); a book on the poet E. E. Cummings; a book on Herbert Hoover; a book on the American circus; and approximately fifty articles on literature, history, and popular culture. He has lectured in Canada, England, Germany, Italy, Russia, Poland, the Czech Republic, Rumania, Austria, Malaysia, and Thailand, as well as across the United States. He was a John Hay Whitney Fellow at Yale University in 1961–62 and a National Endowment for the Humanities Fellow at the University of Illinois (summer 1986) and the University of Wisconsin (summer 1990). In 1987 (summer), he received a grant from the United States Army to study military history at West Point. Eckley holds a Ph.D. from Case Western Reserve University; an M.A. from the Pennsylvania State University; and an A.B. from Mount Union College.

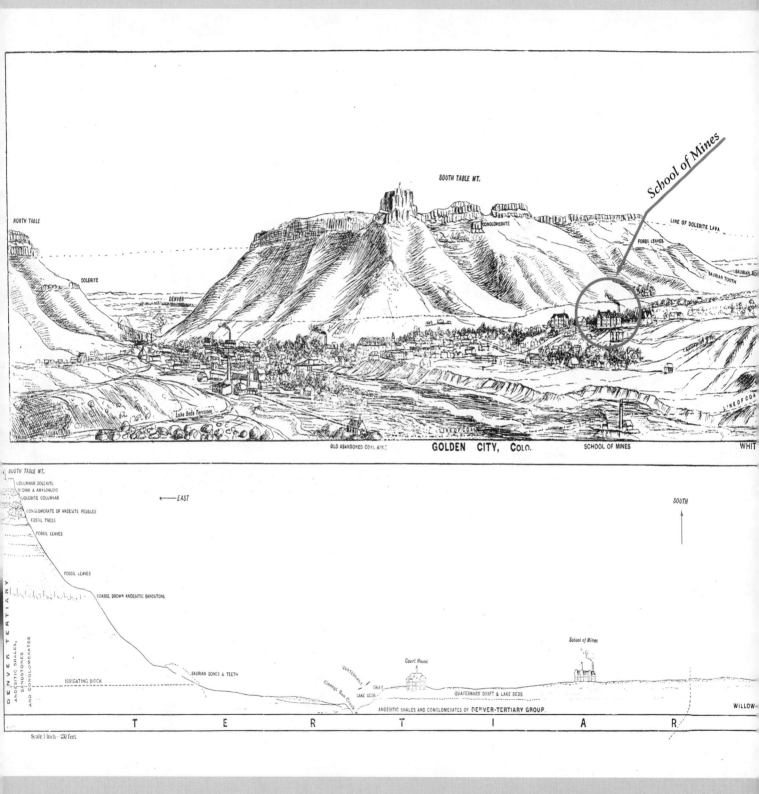